APPLICATIONS OF NONLINEAR CONTROL

Edited by **Meral Altınay**

Applications of Nonlinear Control
Edited by Meral Altınay

Published by InTech

Janeza Trdine 9, 51000 Rijeka, Croatia

Publishing Process Manager Anja Filipovic

Technical Editor Teodora Smiljanic

Cover Designer InTech Design Team

Additional hard copies can be obtained from orders@intechopen.com

Applications of Nonlinear Control, Edited by Meral Altınay
p. cm.
ISBN 978-953-51-0656-2

Meet the editor

Dr Meral Altınay was born in Kocaeli, Turkey, in 1972. She received the M.Sc. and Ph.D. degrees from Kocaeli University, Kocaeli, Turkey, in 1998 and 2005. Since 2006, she has been an Assistant Professor with the Faculty of Technical Education, Kocaeli University. Her major fields of interest include modeling and control of power converters, power quality and nonlinear control.

Contents

Preface

All practical systems contain nonlinear dynamics. Control system development for these systems has traditionally been based on linearized system dynamics in conjunction with linear control techniques. Sometimes it is possible to describe the operation of systems by a linear model around its operating points. Linearized system can provide approximate behavior of the system. But in analyzing the overall system behavior, the resulting system model is inadequate or inaccurate. Moreover, the stability of the system cannot be guaranteed. However, nonlinear control techniques take advantage of the given nonlinear dynamics to produce high-performance designs.

Nonlinear Control Systems represent a new trend of investigation during the last few decades. There has been great excitement over the development of new mathematical techniques for the control of nonlinear systems. Methods for the analysis and design of nonlinear control systems have improved rapidly. A number of new approaches, ideas and results have emerged during this time. These developments have been motivated by comprehensive applications such as mechatronic, robotics, automotive and air-craft control systems.

The book is organized into eleven chapters that include nonlinear design topics such as Feedback Linearization, Lyapunov Based Control, Adaptive Control, Optimal Control and Robust Control. All chapters discuss different applications that are basically independent of each other. The book will provide the reader with information on modern control techniques and results which cover a very wide application area. Each chapter attempts to demonstrate how one would apply these techniques to real-world systems through both simulations and experimental settings.

Lastly, I would like to thank all the authors for their excellent contributions in different applications of Nonlinear Control Techniques. Despite the rapid advances in the field, I believe that the examples provided here allow us to look through some main research tendencies in the upcoming years. I hope the book will be a worthy contribution to the field of Nonlinear Control, and hopefully it will provide the readers with different points of view on this interesting branch of Control Engineering.

Dr. Meral Altınay
Kocaeli University,
Turkey

1

Application of Input-Output Linearization

Erdal Şehirli and Meral Altinay
Kastamonu University & Kocaeli University
Turkey

1. Introduction

In nature, most of the systems are nonlinear. But, most of them are thought as linear and the control structures are realized with linear approach. Because, linear control methods are so strong to define the stability of the systems. However, linear control gives poor results in large operation range and the effects of hard nonlinearities cannot be derived from linear methods. Furthermore, designing linear controller, there must not be uncertainties on the parameters of system model because this causes performance degradation or instability. For that reasons, the nonlinear control are chosen. Nonlinear control methods also provide simplicity of the controller (Slotine & Li, 1991).

There are lots of machine in industry. One of the basic one is dc machine. There are two kinds of dc machines which are brushless and brushed. Brushed type of dc machine needs more maintenance than the other type due to its brush and commutator. However, the control of brushless dc motor is more complicated. Whereas, the control of brushed dc machine is easier than all the other kind of machines. Furthermore, dc machines need to dc current. This dc current can be supplied by dc source or rectified ac source. Three - phase ac source can provide higher voltage than one phase ac source. When the rectified dc current is used, the dc machine can generate harmonic distortion and reactive power on grid side. Also for the speed control, the dc source must be variable. In this paper, dc machine is fed by three - phase voltage source pulse width modulation (PWM) rectifier. This kind of rectifiers compared to phase controlled rectifiers have lots of advantages such as lower line currents harmonics, sinusoidal line currents, controllable power factor and dc - link voltage. To make use of these advantages, the filters that are used for grid connection and the control algorithm must be chosen carefully.

In the literature there are lots of control methods for both voltage source rectifier and dc machine. References (Ooi et al., 1987; Dixon&Ooi, 1988; Dixon, 1990; Wu et al., 1988, 1991) realize current control of L filtered PWM rectifier at three - phase system. Reference (Blasko & Kaura, 1997) derives mathematical model of Voltage Source Converter (VSC) in d-q and α-β frames and also controlled it in d-q frames, as in (Bose, 2002; Kazmierkowski et al., 2002). Reference (Dai et al., 2001) realizes control of L filtered VSC with different decoupling structures. The design and control of LCL filtered VSC are carried out in d-q frames, as in (Lindgren, 1998; Liserre et al., 2005; Dannehl et al., 2007). Reference (Lee et al., 2000; Lee, 2003) realize input-output nonlinear control of L filtered VSC, and also in reference (Kömürcügil & Kükrer, 1998) Lyapunov based controller is designed for VSC. The feedback linearization technique for LCL filtered VSC is also presented, as in (Kim & Lee, 2007; Sehirli

& Altınay, 2010). Reference (Holtz, 1994) compares the performance of pulse width modulation (PWM) techniques which are used for VSC. In (Krishnan, 2001) the control algorithms, theories and the structure of machines are described. The fuzzy and neural network controls are applied to dc machine, as in (Bates et al., 1993; Sousa & Bose, 1994).

In this chapter, simulation of dc machine speed control which is fed by three - phase voltage source rectifier under input - output linearization nonlinear control, is realized. The speed control loop is combined with input-output linearization nonlinear control. By means of the simulation, power factor, line currents harmonic distortions and dc machine speed are presented.

2. Main configuration of VSC

In many industrial applications, it is desired that the rectifiers have the following features; high-unity power factor, low input current harmonic distortion, variable dc output voltage and occasionally, reversibility. Rectifiers with diodes and thyristors cannot meet most of these requirements. However, PWM rectifiers can provide these specifications in comparison with phase-controlled rectifiers that include diodes and thyristors.

The power circuit of VSC topology shown in Fig.1 is composed of six controlled switches and input L filters. Ac-side inputs are ideal three-phase symmetrical voltage source, which are filtered by inductor L and parasitic resistance R, then connected to three-phase rectifier consist of six insulated gate bipolar transistors (IGBTs) and diodes in reversed parallel. The output is composed of capacitance and resistance.

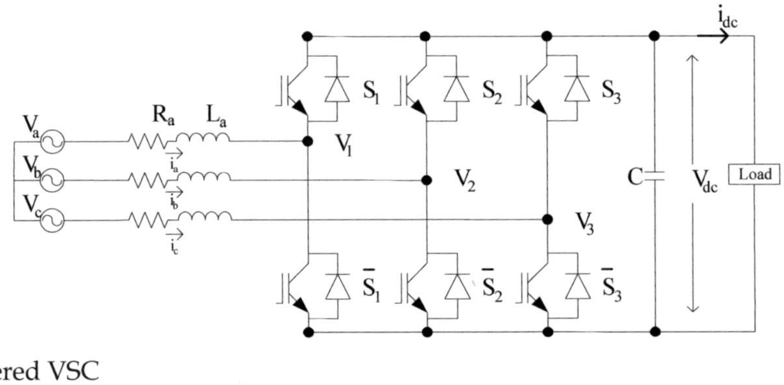

Fig. 1. L filtered VSC

3. Mathematical model of the VSC

3.1 Model of the VSC in the three-phase reference frame

Considering state variables on the circuit of Fig.1 and applying Kirchhoff laws, model of VSC in the three-phase reference frame can be obtained, as in (Wu et al., 1988, 1991; Blasko & Kaura, 1997).

The model of VSC is carried out under the following assumptions.

- The power switches are ideal devices.
- All circuit elements are LTI (Linear Time Invariant)
- The input AC voltage is a balanced three-phase supply.

For the three-phase voltage source rectifier, the phase duty cycles are defined as the duty cycle of the top switch in that phase, i.e., $d_a = d(S_1)$, $d_b = d(S_3)$, $d_c = d(S_5)$ with d representing duty cycle.

$$\frac{di_a}{dt} = -\frac{R}{L}i_a - V_{dc}\left(d_a - \frac{1}{3}\sum_{k=a}^{c} d_k\right) + V_a \quad (1)$$

$$\frac{di_b}{dt} = -\frac{R}{L}i_b - V_{dc}\left(d_b - \frac{1}{3}\sum_{k=a}^{c} d_k\right) + V_b \quad (2)$$

$$\frac{di_c}{dt} = -\frac{R}{L}i_c - V_{dc}\left(d_c - \frac{1}{3}\sum_{k=a}^{c} d_k\right) + V_c \quad (3)$$

$$\frac{dV_{dc}}{dt} = \frac{1}{C}(i_a d_a + i_b d_b + i_c d_c) - \frac{1}{C}i_{dc} \quad (4)$$

This model in equations (1) – (4) is nonlinear and time variant. Using Park Transformation, the ac-side quantities can be transformed into rotating d-q frame. Therefore, it is possible to obtain a time-invariant system model with a lower order.

3.2 Coordinate transformation

On the control of VSC, to make a transformation, there are three coordinates whose relations are shown by Fig 2, that are a-b-c, α-β and d-q. a-b-c is three phase coordinate, α-β is stationary coordinate and d-q is rotating coordinate which rotates ω speed.

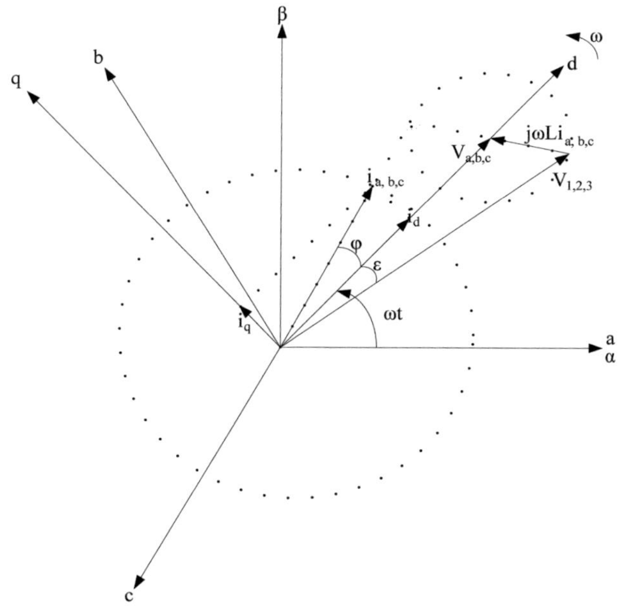

Fig. 2. Coordinates diagram of a-b-c, α-β and d-q

The d-q transformation is a transformation of coordinates from the three-phase stationary coordinate system to the d-q rotating coordinate system. A representation of a vector in any n-dimensional space is accomplished through the product of a transpose n-dimensional vector (base) of coordinate units and a vector representation of the vector, whose elements are corresponding projections on each coordinate axis, normalized by their unit values. In three phase (three dimensional) space, it looks like as in (5).

$$X_{abc} = \begin{bmatrix} a_u & b_u & c_u \end{bmatrix} \begin{bmatrix} x_a \\ x_b \\ x_c \end{bmatrix} \tag{5}$$

Assuming a balanced three-phase system, a three-phase vector representation transforms to d-q vector representation (zero-axis component is 0) through the transformation matrix T, defined as in (6).

$$T = \frac{2}{3} \begin{bmatrix} \cos(\omega t) & \cos(\omega t - \frac{2}{3}\pi) & \cos(\omega t + \frac{2}{3}\pi) \\ -\sin(\omega t) & -\sin(\omega t - \frac{2}{3}\pi) & -\sin(\omega t + \frac{2}{3}\pi) \end{bmatrix} \tag{6}$$

In (6), ω is the fundamental frequency of three-phase variables. The transformation from X_{abc} (three-phase coordinates) to X_{dq}(d-q rotating coordinates), called Park Transformation, is obtained through the multiplication of the vector X_{abc} by the matrix T, as in (7).

$$X_{dq} = T.X_{abc} \tag{7}$$

The inverse transformation matrix (from d-q to a-b-c) is defined in (8).

$$T' = \frac{2}{3} \begin{bmatrix} \cos(\omega t) & -\sin(\omega t + \frac{2}{3}\pi) \\ \cos(\omega t - \frac{2}{3}\pi) & -\sin(\omega t - \frac{2}{3}\pi) \\ \cos(\omega t + \frac{2}{3}\pi) & -\sin(\omega t + \frac{2}{3}\pi) \end{bmatrix} \tag{8}$$

The inverse transformation is calculated as in (9).

$$X_{abc} = T'.X_{dq} \tag{9}$$

3.3 Model of the VSC in the rotating frame

Let x and u be the phase state variable vector and phase input vector in one phase of a balanced three-phase system with the state equation in one phase as in (10).

$$\dot{X} = Ax + Bu \tag{10}$$

Where A and B are identical for the three phases. Applying d-q transformation to the three-phase system, d-q subsystem with d and q variables is obtained (x_d-x_q and u_d-u_q). The system equation in (10) becomes as in (11) (Mao et al., 1998; Mihailovic, 1998).

$$\begin{bmatrix} \dot{X}_d \\ \dot{X}_q \end{bmatrix} = \begin{bmatrix} A & \omega I \\ -\omega I & A \end{bmatrix} \begin{bmatrix} x_d \\ x_q \end{bmatrix} + \begin{bmatrix} B & 0 \\ 0 & B \end{bmatrix} \begin{bmatrix} u_d \\ u_q \end{bmatrix} \tag{11}$$

Where I is the identity matrix and 0 is a zero matrix, both having the same dimension as x. (11) can transform any three-phase system into the d-q model directly.

When equations (1) – (4) are transformed into d-q coordinates, (12) – (14) are obtained, as in (Blasko & Kaura, 1997; Ye, 2000; Kazmierkowski et al., 2002).

$$\frac{di_d}{dt} = -\frac{R}{L}i_d + \omega i_q - \frac{1}{L}V_{dc}d_d - \frac{U_d}{L} \tag{12}$$

$$\frac{di_q}{dt} = -\frac{R}{L}i_q - \omega i_d - \frac{1}{L}V_{dc}d_q - \frac{U_q}{L} \tag{13}$$

$$\frac{dV_{dc}}{dt} = \frac{1}{C}(i_d d_d + i_q d_q) - \frac{1}{C}i_{dc} \tag{14}$$

Where i_d and i_q are the d-q transformation of i_a, i_b and i_c. v_d and v_q are the d-q transformation of v_a , v_b and v_c. d_d and d_q are the d-q transformation of d_a, d_b and d_c .

4. Input-output feedback linearization technique

Feedback linearization can be used as a nonlinear design methodology. The basic idea is first to transform a nonlinear system into a (fully or partially) linear system, and then to use the well-known and powerful linear design techniques to complete the control design. It is completely different from conventional linearization. In feedback linearization, instead of linear approximations of the dynamics, the process is carried out by exact state transformation and feedback. Besides, it is thought that the original system is transformed into an equivalent simpler form. Furthermore, there are two feedback linearization methods that are input-state and input-output feedback linearization (Slotine & Li, 1991; Isidori, 1995; Khalil, 2000; Lee et al., 2000; Lee, 2003).

The input-output feedback linearization technique is summarized by three rules;

- Deriving output until input appears
- Choosing a new control variable which provides to reduce the tracking error and to eliminate the nonlinearity
- Studying stability of the internal dynamics which are the part of system dynamics cannot be observed in input-output linearization (Slotine & Li, 1991)

If it is considered an input-output system, as in (15)-(16);

$$\dot{X} = f(x) + g(x)u \tag{15}$$

$$y = h(x) \tag{16}$$

To obtain input-output linearization of this system, the outputs y must be differentiated until inputs u appears. By differentiating (16), equation (17) is obtained.

$$\dot{y} = \frac{\partial h}{\partial x}[f(x) + g(x)u] = L_f h(x) + L_g h(x)u \tag{17}$$

In (17), $L_f h$ and $L_g h$ are the Lie derivatives of f(x) and h(x), respectively and identified in (18).

$$L_f h(x) = \frac{\partial h}{\partial x}f(x), \quad L_g h(x) = \frac{\partial h}{\partial x}g(x) \tag{18}$$

If the k is taken as a constant value; k. order derivatives of h(x) and 0. order derivative of h(x) are shown in (19) - (20), respectively.

$$L_f^k h(x) = L_f L_f^{k-1} h(x) = \frac{\partial (L_f^{k-1} h)}{\partial x} f(x) \tag{19}$$

$$L_f^0 h(x) = h(x) \tag{20}$$

After first derivation, If $L_g h$ is equal to "0", the output equation becomes $\dot{y} = L_f h(x)$. However, it is independent from u input. Therefore, it is required to take a derivative of output again. Second derivation of output can be written in (23), with the help of (21)-(22).

$$L_g L_f h(x) = \frac{\partial (L_f h)}{\partial x} g(x) \tag{21}$$

$$L_f^2 h(x) = L_f L_f h(x) = \frac{\partial (L_f h)}{\partial x} f(x) \tag{22}$$

$$\ddot{y} = \frac{\partial L_f h}{\partial x} [f(x) + g(x)u] = L_f^2 h(x) + L_g L_f h(x) u \tag{23}$$

If $L_g L_f h(x)$ is again equal to "0", $\ddot{y}$ is equal to $L_f^2 h(x)$and it is also independent from u input and it is continued to take the derivation of output. After r times derivation, if the condition of (24) is provided, input appears in output and (25) is obtained.

$$L_{g_i} L_f^{r_i - 1} h_i(x) \neq 0 \tag{24}$$

$$y_i^{r_i} = L_f^{r_i} h_i + \sum_{i=1}^{n} \left(L_{g_i} L_f^{r_i - 1} h_i \right) u_i \tag{25}$$

Applying (25) for all n outputs, (26) is derived.

$$\begin{bmatrix} y_1^{r_1} \\ \cdots \\ \cdots \\ y_1^{r_n} \end{bmatrix} = \begin{bmatrix} L_f^{r_1} h_1(x) \\ \cdots \\ \cdots \\ L_f^{r_n} h_n(x) \end{bmatrix} + E(x) \begin{bmatrix} u_1 \\ \cdots \\ \cdots \\ u_n \end{bmatrix} = a(x) + E(x)u \tag{26}$$

E(x) in (27) is a decoupling matrix, if it is invertible and new control variable is chosen, feedback transformation is obtained, as in (28).

$$E(x) = \begin{bmatrix} L_{g_1} L_f^{r_1 - 1} h_1 & \cdots & L_{g_n} L_f^{r_n - 1} h_1 \\ \vdots & \ddots & \vdots \\ L_{g_1} L_f^{r_n - 1} & \cdots & L_{g_n} L_f^{r_n - 1} h_n \end{bmatrix} \tag{27}$$

$$\begin{bmatrix} u_1 \\ \cdots \\ \cdots \\ u_n \end{bmatrix} = -E^{-1} \begin{bmatrix} L_f^{r_1} h_1(x) \\ \cdots \\ \cdots \\ L_f^{r_n} h_n(x) \end{bmatrix} + E^{-1} \begin{bmatrix} v_1 \\ \cdots \\ \cdots \\ v_n \end{bmatrix} \tag{28}$$

Equation (29) shows the relation between the new inputs v and the outputs y. The input-output relation is decoupled and linear (Lee et al., 2000).

$$\begin{bmatrix} y_1^{r_1} \\ \cdots \\ \cdots \\ y_1^{r_n} \end{bmatrix} = \begin{bmatrix} v_1 \\ \cdots \\ \cdots \\ v_n \end{bmatrix} \quad (29)$$

If the closed loop error dynamics is considered, as in (30) - (31), (32) defines new inputs for tracking control.

$$\begin{bmatrix} e_1^r + k_{1(r-2)}e_1^{r-1} + \cdots + k_{11}e_1^1 + k_{10}e_1 \\ \cdots \\ \cdots \\ e_n^r + k_{n(r-1)}e_n^{r-1} + \cdots + k_{21}e_1^1 + k_{20}e_n \end{bmatrix} = \begin{bmatrix} 0 \\ \cdots \\ \cdots \\ 0 \end{bmatrix} \quad (30)$$

$$\begin{bmatrix} e \\ \cdots \\ \cdots \\ e^r \end{bmatrix} = \begin{bmatrix} y - y^* \\ \cdots \\ \cdots \\ y^r - y^{*^r} \end{bmatrix} \quad (31)$$

$$\begin{bmatrix} v_1 \\ \cdots \\ \cdots \\ v_n \end{bmatrix} = \begin{bmatrix} -k_{1(r-1)}y^{r-1} - \cdots - k_{11(r-1)}y^1 - k_{10}(y_1 - y_1^*) \\ \cdots \\ \cdots \\ -k_{n(r-1)}y^{r-1} - \cdots - k_{21(r-1)}y^1 - k_{20}(y_n - y_n^*) \end{bmatrix} \quad (32)$$

k values in equations show the constant values for stability of systems and tracking of y references, as in (Lee, 2003).

5. The application of an input-output feedback linearization to the VSC

The state feedback transformation allows the linear and independent control of the d and q components of the line currents in VSC by means of the new inputs u_d and u_q .

For unity power factor, in (12 - 14) $u_d = V_m$ and $u_q = 0$ are taken, so mathematical model of this system is derived with (33-35), as in (Kömürcügil & Kükrer, 1998; Lee, 2003).

$$\frac{di_d}{dt} = -\frac{R}{L}i_d + \omega i_q - \frac{1}{L}V_{dc}d_d - \frac{V_m}{L} \quad (33)$$

$$\frac{di_q}{dt} = -\frac{R}{L}i_q - \omega i_d - \frac{1}{L}V_{dc}d_q \quad (34)$$

$$\frac{dV_{dc}}{dt} = \frac{1}{C}(i_d d_d + i_q d_q) - \frac{1}{C}i_{dc} \quad (35)$$

If (33-35) are written with the form of (15), (36) is derived.

$$f(x) = \begin{bmatrix} -\frac{R}{L}\, i_d + \omega i_q + \frac{V_m}{L} \\ -\omega i_d - \frac{R}{L} i_q \\ \frac{1}{C} i_{dc} \end{bmatrix}, \; g(x) = \begin{bmatrix} -\frac{1}{L}V_{dc} & 0 \\ 0 & -\frac{1}{L}V_{dc} \\ \frac{1}{C}i_d & \frac{1}{C}i_q \end{bmatrix} \quad (36)$$

The main purpose of this control method is to regulate V_{dc} voltage by setting i_d current and to provide unity power factor by controlling i_q current. Therefore, variables of y outputs and reference values are chosen as in (37).

$$y = \begin{bmatrix} y_1 \\ y_2 \end{bmatrix} = \begin{bmatrix} h_1(x) \\ h_2(x) \end{bmatrix} = \begin{bmatrix} i_d \\ i_q \end{bmatrix} \;, y^* = \begin{bmatrix} I_d^* \\ 0 \end{bmatrix} \tag{37}$$

Differentiating outputs of (37), (38) is obtained. The order of derivation process, finding a relation between y outputs and u inputs, is called as relative degree. It is also seen that the relative degree of the system is '1'.

$$\dot{y} = \begin{bmatrix} \dot{i_d} \\ \dot{i_q} \end{bmatrix} = \begin{bmatrix} -\frac{1}{L}V_{dc} & 0 \\ 0 & -\frac{1}{L}V_{dc} \end{bmatrix} u + \begin{bmatrix} -\frac{R}{L}i_d + \omega i_q + \frac{V_m}{L} \\ -\omega i_d - \frac{R}{L}i_q \end{bmatrix} \tag{38}$$

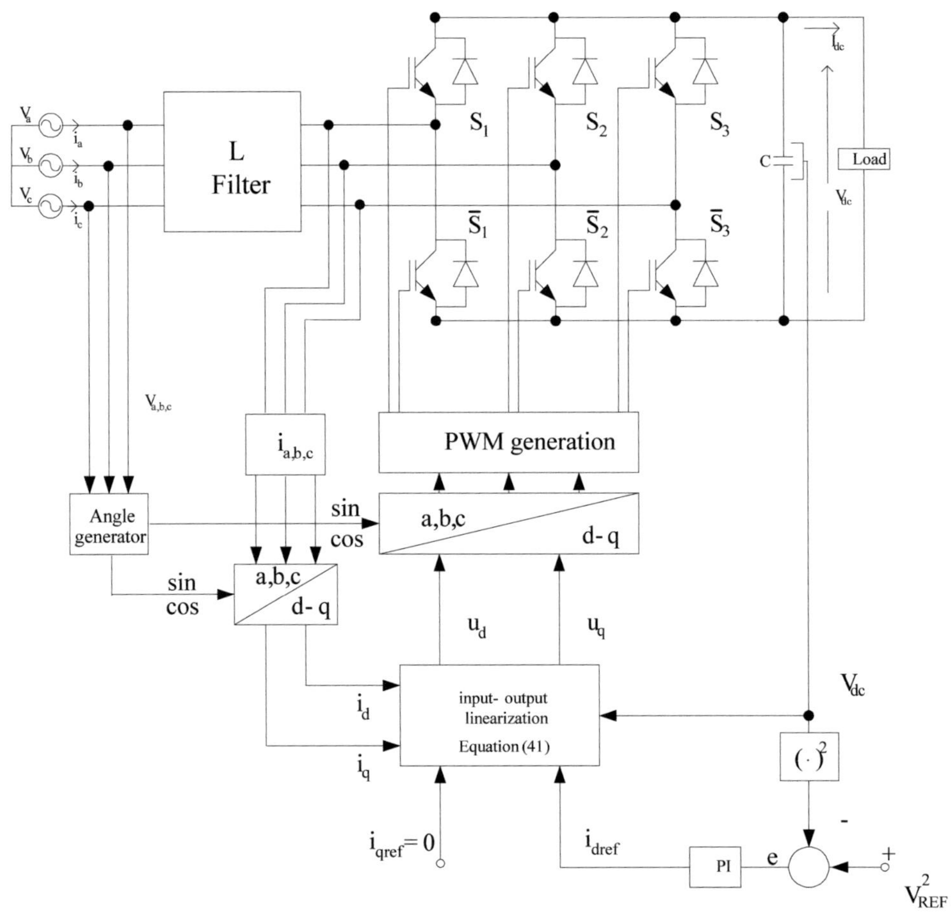

Fig. 3. Input-output feedback linearization control algorithm of VSC

When (38) is ordered like (28), (39) is obtained.

$$u = \begin{bmatrix} u_d \\ u_q \end{bmatrix} = \begin{bmatrix} -\frac{1}{L}V_{dc} & 0 \\ 0 & -\frac{1}{L}V_{dc} \end{bmatrix}^{-1} \cdot \left(-\begin{bmatrix} -\frac{R}{L}i_d + \omega i_q + \frac{V_m}{L} \\ -\omega i_d - \frac{R}{L}i_q \end{bmatrix} + v \right) \tag{39}$$

After taking inverse of matrix (39) and adding new control inputs from (40), (41) is obtained.

$$\begin{bmatrix} v_1 \\ v_2 \end{bmatrix} = \begin{bmatrix} -k_1(y_1 - I_d^*) \\ -k_2(y_2 - I_q^*) \end{bmatrix} \tag{40}$$

$$\begin{bmatrix} u_d \\ u_q \end{bmatrix} = \begin{bmatrix} -\frac{L}{V_{dc}} & 0 \\ 0 & -\frac{L}{V_{dc}} \end{bmatrix} \cdot \left(\begin{bmatrix} \frac{R}{L} i_d - \omega i_q - \frac{V_m}{L} \\ \omega i_d + \frac{R}{L} i_q \end{bmatrix} + \begin{bmatrix} -k_1(y_1 - I_d^*) \\ -k_2 y_2 \end{bmatrix} \right) \tag{41}$$

Control algorithm is seen in Fig.3. For both L and LCL filtered VSC, the same control algorithm can be used. Providing the unity power factor, angle values are obtained from line voltages. This angle values are used in transformation of a-b-c to d-q frames. Line currents which are transformed into d-q frame, are compared with d-q reference currents. d axis reference current i_{dref} is obtained by comparison of dc reference voltage V_{ref} and actual dc voltage V_{dc}. On the other hand, q axis reference current i_{qref} is set to '0' to provide unity power factor. And by using (41), switching functions of d-q components are found. Then, this d-q switching functions are transformed into a-b-c frames and they are sent to PWM block to produce pulses for power switches.

Providing the control of internal dynamics, in dc controller square of V_{ref} and V_{dc} are used (Lee, 2003).

6. DC machine and armature circuit model

Electrical machines are used for the conversion of electric power to mechanical power or vice versa. In industry, there are wide range of electrical machines that are dc machines, induction machines, synchronous machines, variable reluctance machines and stepping motors. The Dc machines can be classified as a brushless and brushed dc machines. Furthermore, the advantage of brushed dc machines is the simplicity with regard to speed control in the whole machines. However, the main disadvantage of this kind of machines is the need of maintenance because of its brushes and commutators.

Fig. 4 shows the basic structure of brushed dc machines. Basic components of dc machines are field poles, armature, commutator and brushes (Fitzgerald et al., 2003).

Fig. 4. Dc machine

Field poles produce the main magnetic flux inside of the machines with the help of the field coils which are wound around the poles and carry the field current. Some of the dc machines, the magnetic flux is provided by the permanent magnet instead of the field coils. In Fig. 5, the field coils and field poles of dc machines are shown (Bal, 2008).

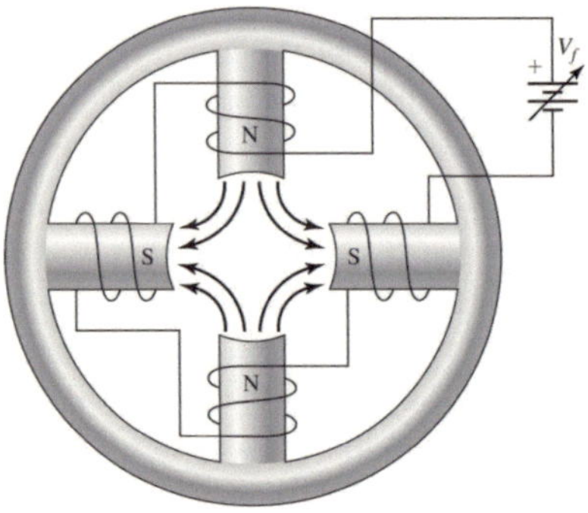

Fig. 5. Field coils and field poles of a dc machine

The rotating part of the dc machine is called as an armature. The armature consists of iron core, conductors and commutator. Besides, there is a shaft inside of armature that rotates between the field poles. The other part of the machine is commutator which is made up of copper segments and it is mounted on the shaft. Furthermore, the armature conductors are connected on the commutator. Another component of dc machine is brushes. The brushes provide the electric current required by armature conductors. In dc machine to ensure the rotation of the shaft, the armature conductors must be energized. This task is achieved by brushes that contact copper segments of commutator. Also, the brushes generally consist of carbon due to its good characteristic of electrical permeability. Fig. 6 shows the armature, commutator and the brushes (Fitzgerald et al., 2003; Bal, 2008).

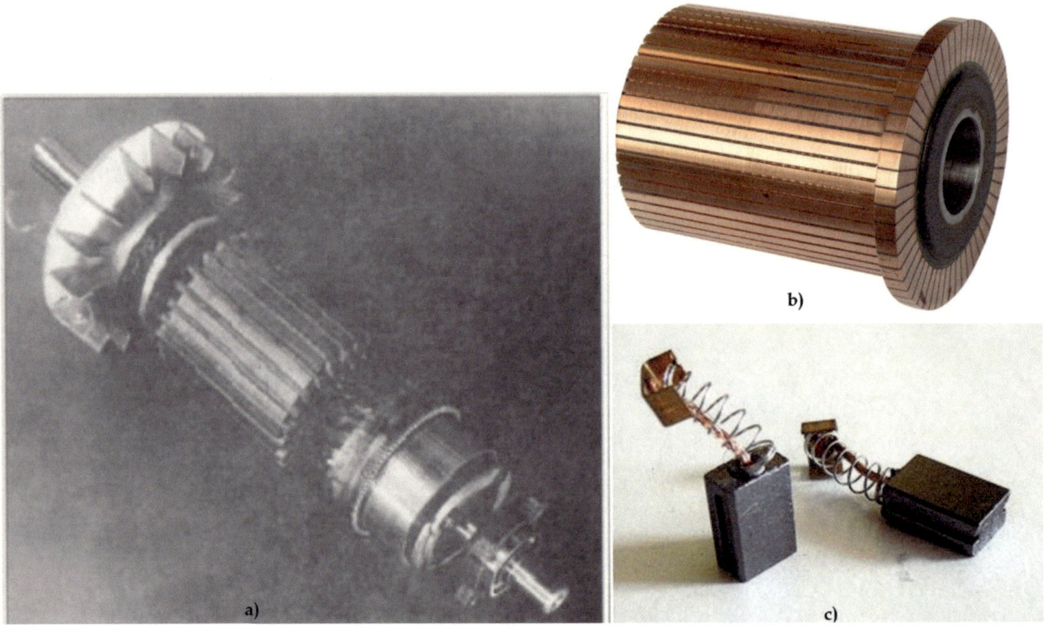

Fig. 6. a) armature, b) commutator and c) brushes of a dc machine

To produce the main flux, the field must be excited. For this task, there are four methods which are separately, shunt, series and compound to excitation of dc machines and are shown in Fig. 7. However, separately excited dc machine is the most useful method because it provides independent control of field and armature currents. Therefore, this structure is used in this chapter (Krishnan, 2001; Fitzgerald et al., 2003).

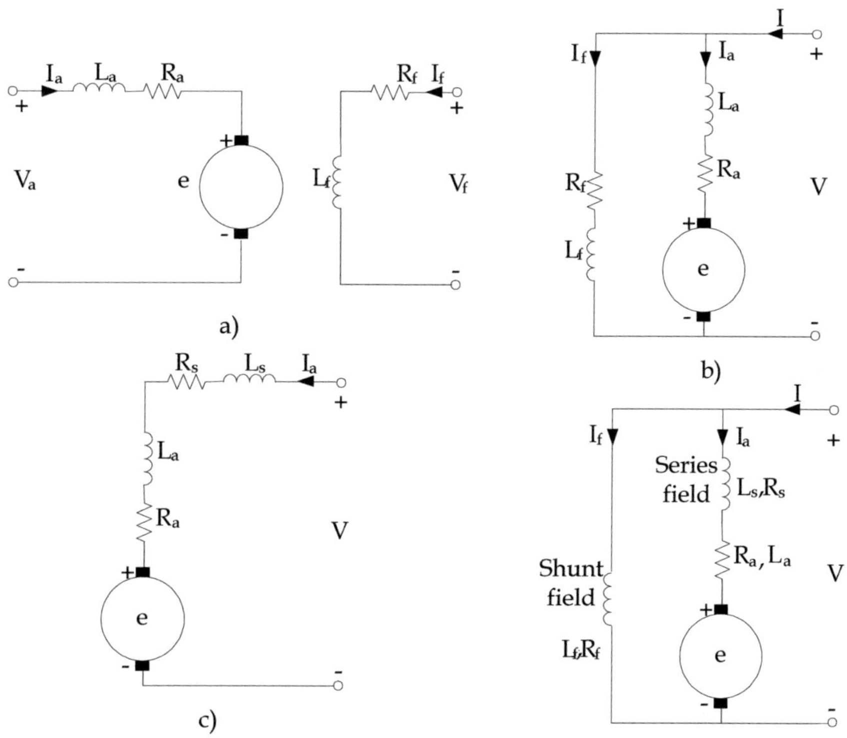

Fig. 7. Excitation methods of dc machine a) separately, b) shunt, c) series, d) compound excitation

There are two basic speed control structure of dc machine which are armature and field, as in (Krishnan, 2001). The armature circuit model of dc machine is shown in Fig. 8.

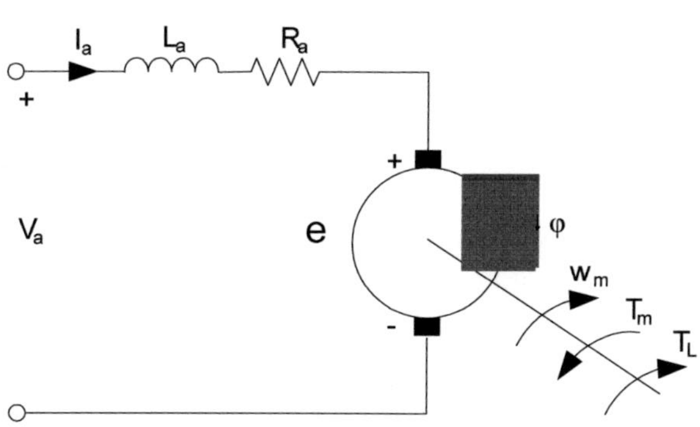

Fig. 8. Armature circuit model of dc machine

The mathematical model of armature circuit can be written by (42).

$$v = e + R_a I_a + L_a \frac{di_a}{dt} \tag{42}$$

In steady state, $\frac{di_a}{dt}$ part is zero because of the armature current is constant. The armature model is then obtained by (43), (Krishnan, 2001).

$$v = e + R_a I_a \tag{43}$$

$$e = K\Phi_f \omega_m \tag{44}$$

(43) is written in (44), (45) is derived.

$$\omega_m \approx \frac{(V - R_a I_a)}{I_f} \tag{45}$$

The speed of dc machine depends on armature voltage and field current, as shown in (45). In field control, the armature voltage is kept constant and the field current is set. The relation between speed and field current is indirect proportion. However, in armature control, the relation between armature voltage and speed is directly proportional. Furthermore, in armature control, the field current is kept constant and the armature voltage is set.

In this chapter, the armature control of dc machine is realized.

The speed control loop is added to nonlinear control loop. Firstly, the actual speed is compared with reference speed then the speed error is regulated by PI controller and after that its subtraction from armature current, the reference current is obtained. The reference current obtained by speed loop, is added to nonlinear control loop instead of reference i_d current, which is obtained by the comparison of the square of reference voltage and actual voltage.

In Fig. 9, speed control loop is shown.

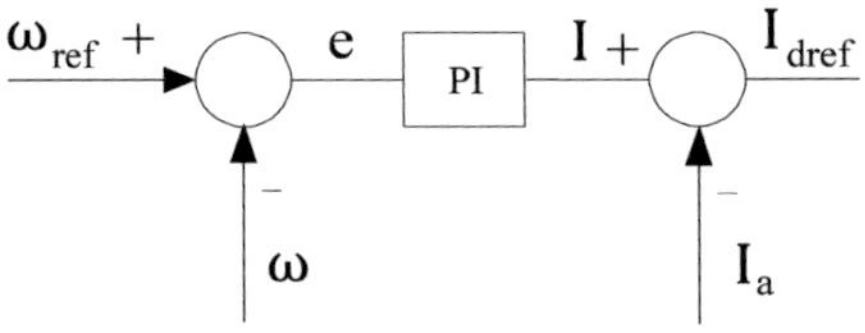

Fig. 9. Armature speed control loop of dc machine

7. Simulations

Simulations are realized with Matlab/Simulink. Line voltage is taken 220 V, 60 Hz. The switching frequency is also chosen 9 kHz. L filter and controllers parameters are shown in Table.1 and Table.2.

Simulation diagram is shown in Fig.10. By simulation, steady-state error and settling time of dc motor speed, harmonic distortions and shapes of line currents and unity power factor are examined.

Passive Components		
L Filter		Dc-Link
L (H)	R (Ω)	C_{dc} (μF)
0.0045	5.5	2200

Table 1. Values of L filter components

Controllers			
Speed Controller		Input-output current controller	
K_p	K_i	$k\ (10^3)$	$K\ (10^5)$
10	0.01	30	50

Table 2. Values of controllers

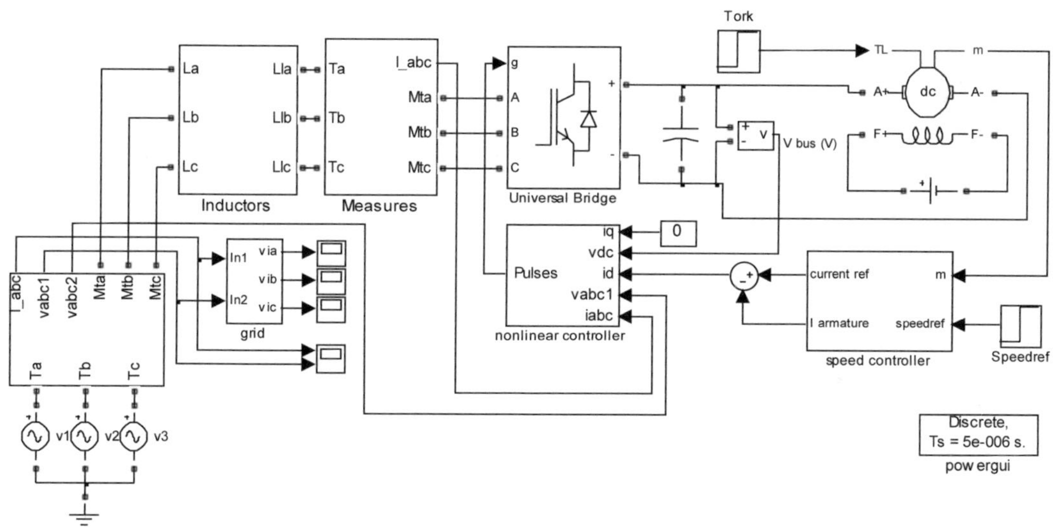

Fig. 10. Simulation diagram of dc machine controller in Simulink

Fig.11 shows the structure of input-output controller diagram.

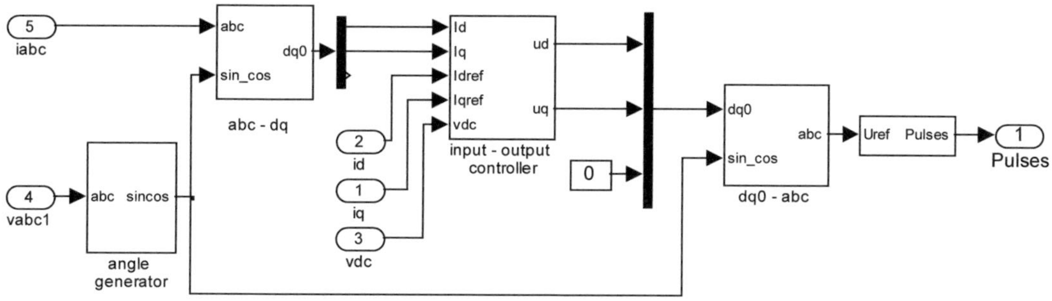

Fig. 11. Input-output controller diagram

Equation (41) is written in the block of input-output controller which is shown in Fig.12.

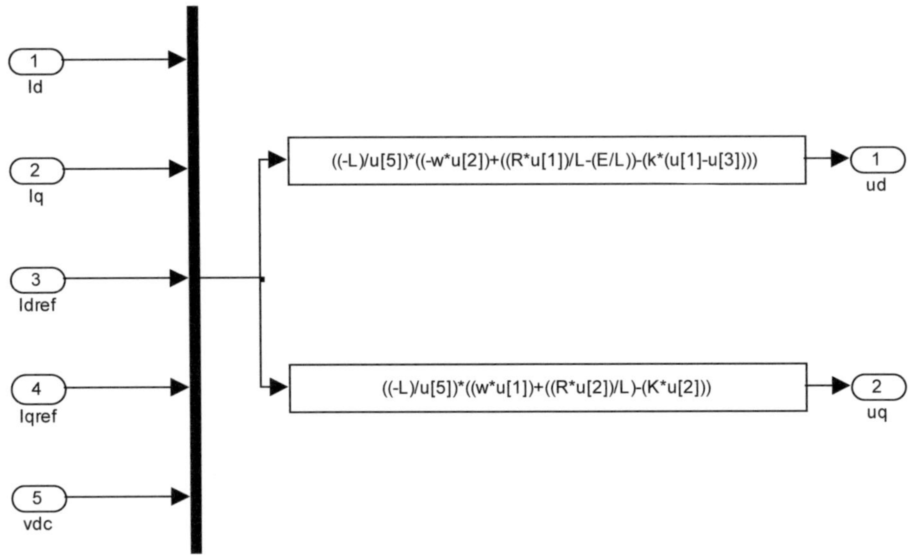

Fig. 12. Input - output controller

Fig. 13 shows the speed controller of dc machine.

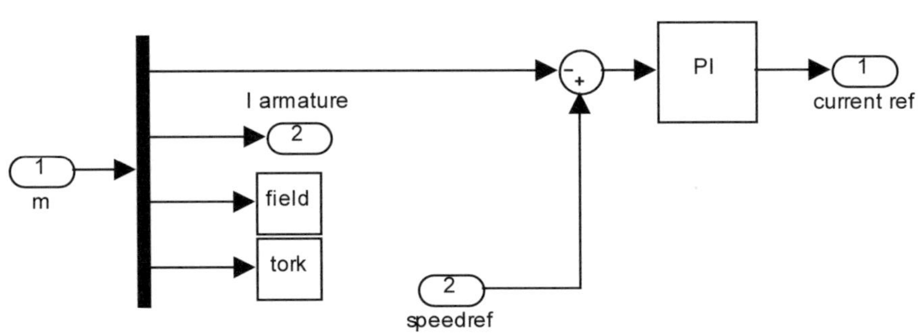

Fig. 13. Speed controller of dc machine

Fig.14 shows the dc machine speed. Reference speed value is changed from 150 rad/s to 200rad/s at 0.5 s. Settling time to the first reference is shorter than 0.15 s, but settling time of second reference is 0.1 s.

Fig. 15 shows the steady-state error of dc machine speed. It is seen that the steady - state error changes between ±2 rad/s.

The one phase voltage and current is shown in Fig. 16. It is also seen that unity power factor is obtained but not as desired.

Fig. 17 shows the line currents. The shapes of line currents are sinusoidal.

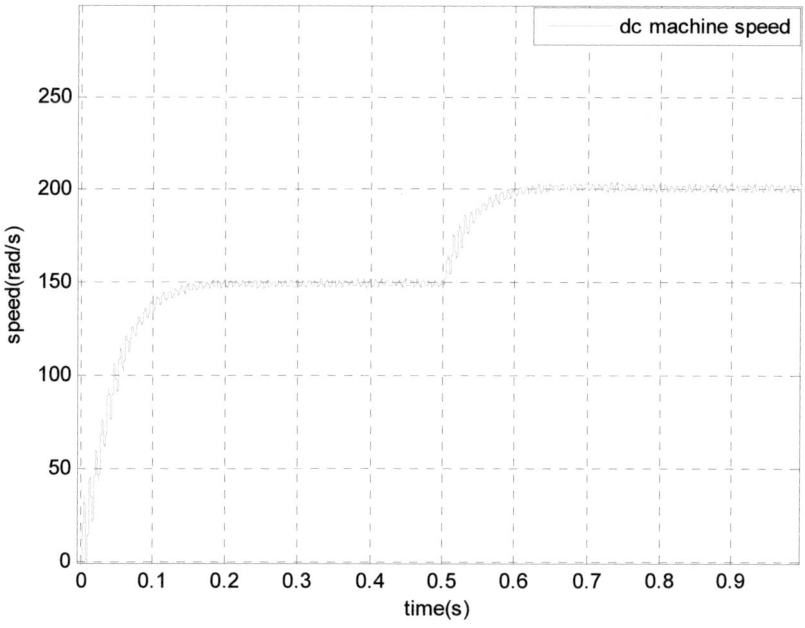

Fig. 14. Dc machine speed

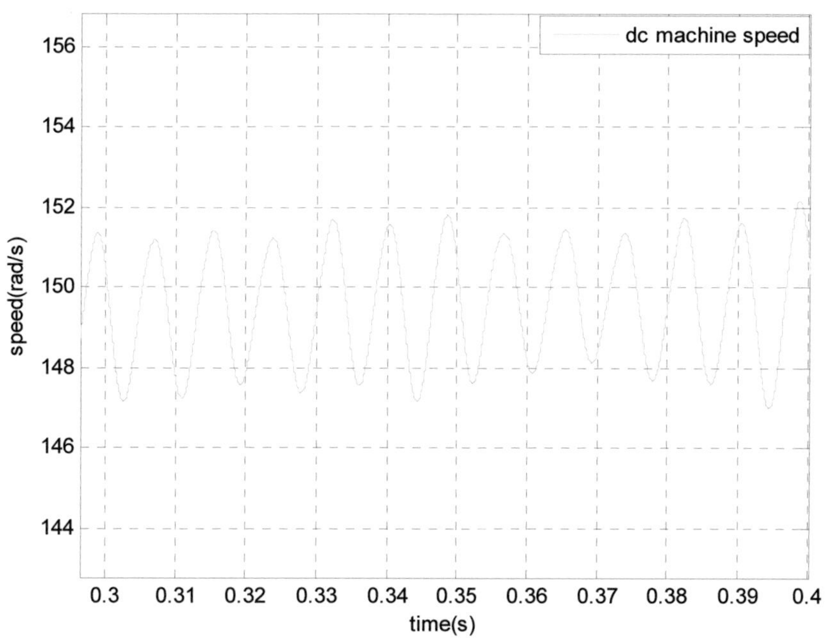

Fig. 15. Steady-state error of dc machine speed

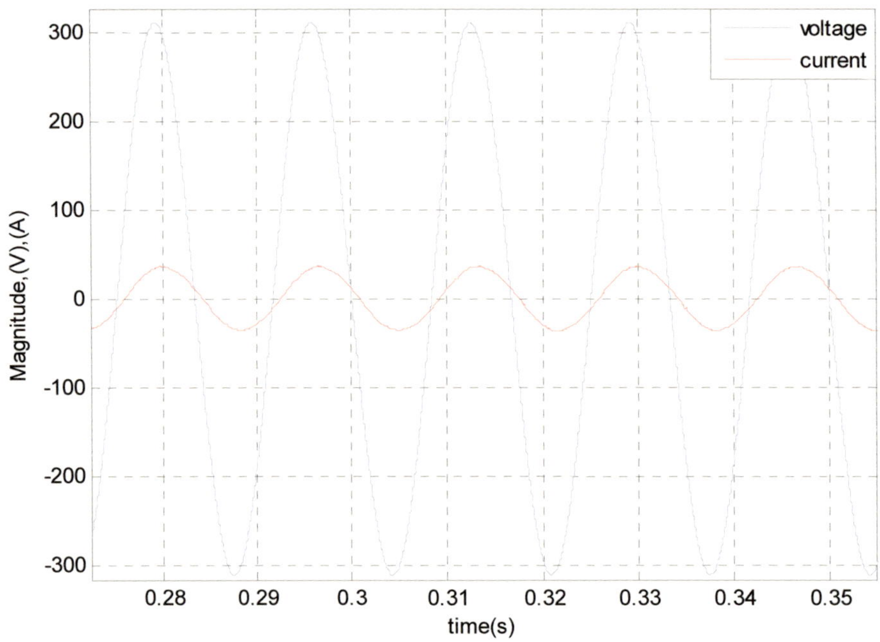

Fig. 16. One phase voltage and current

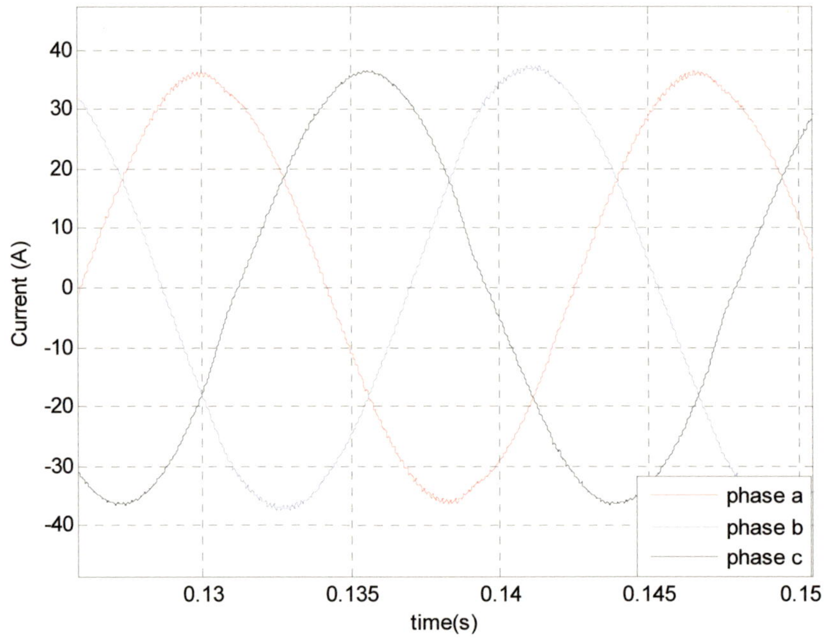

Fig. 17. Three-phase phase current

Fig. 18 shows the harmonic distortions of line currents. Line currents include high order harmonic contents. However, total harmonic distortion value (THD) is under the value that is defined by standards. THD of line currents are %1.34, %1.71 and %2.84.

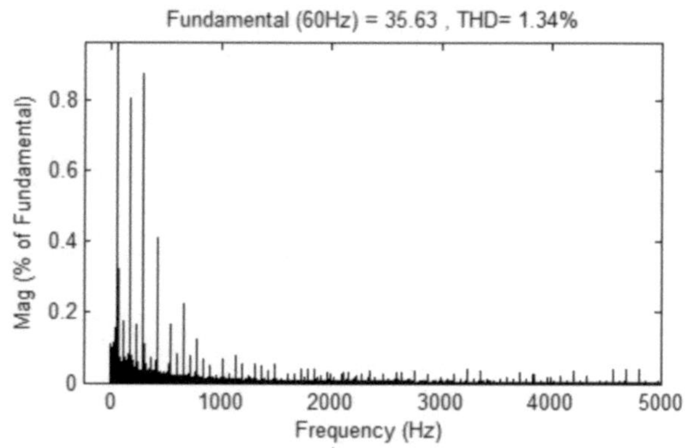

I_a

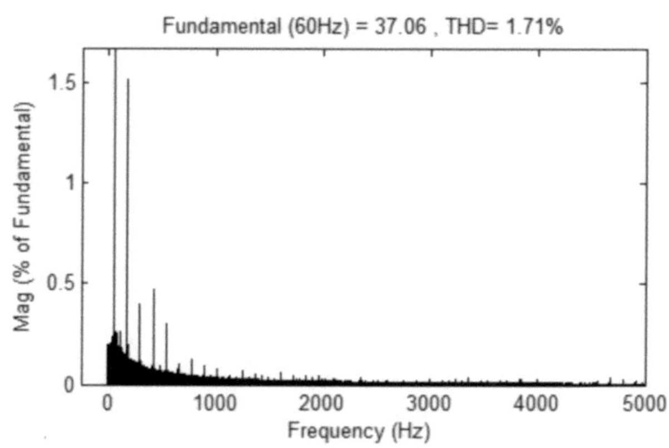

I_b

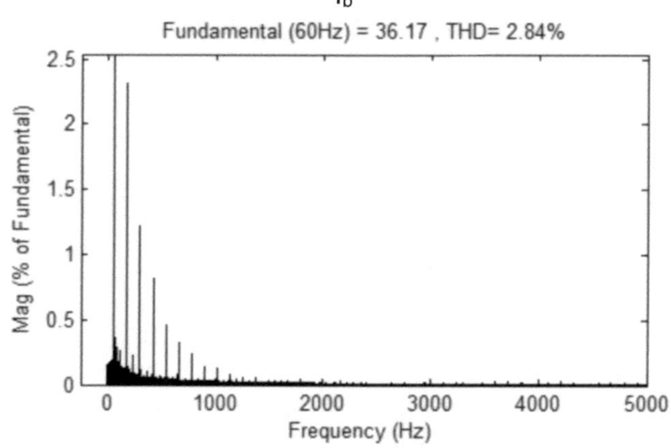

I_c

Fig. 18. Harmonic distortions of line currents

8. Conclusion

In this chapter, simulation of dc machine armature speed control is realized. Dc machine is fed by voltage source rectifier which is controlled input - output linearization nonlinear control method. Furthermore, for the speed control, dc link voltage is regulated by the dc machine speed control loop. The control algorithm of voltage source rectifier and dc motor speed are combined. The required reference I_d current for voltage source rectifier is obtained by speed control loop. Simulations are carried through Matlab/Simulink. By means of the simulation results, the speed of dc machine, line currents harmonic distortions and power factor of grid are shown. It is shown that the voltage source rectifier with dc machine as a load provides lower harmonic distortion and higher power factor. Furthermore, dc machine speed can be regulated.

9. References

Bal, G. (2008). *Dogru Akım Makinaları ve Suruculeri,* Seckin, ISBN 978-975-02-0706-8, Sihhiye, Ankara, Turkey

Bates, J.; Elbuluk, M,E. & Zinger, D,S. (1993). Neural Network Control of a Chopper-Fed Dc Motor, *IEEE Power Electronics Specialist Conference PESC02,* Vol 3, pp. 893, ISBN 0-7803-1243-0, Seattle, Washington, USA, June, 1993

Blasko, V. & Kaura, V. (1997). A New Mathematical Model and Control Three-Phase ac-dc Voltage Source Converter, *IEEE Transactions on Power Electronics,* Vol 12, January, 1997, pp. 78-81, ISSN 0885-8993

Bose, B,K. (2002). *Modern Power Electronics and Ac Drives,* Prentice Hall, ISBN 0-13-016743-6, New Jersey, USA

Dai, K.; Liu, P. ; Kang, Y. & Chen, J. (2001). Decoupling Current Control for Voltage Source Converter in Synchronous Rotating Frame, *IEEE PEDS,* pp. 39-43, ISBN 0-7803-7233-6, Indonesia, January-February, 2001

Dannehl, J.; Fuchs, F,W. & Hansen, S. (2007). PWM Rectifier with LCL filter Using Different Control Structures, *EPE Aalborg,* pp. 1-10, ISBN 972-92-75815-10-2, Aalborg, October, 2007

Dixon, J,W. & Ooi, B,T. (1988). Indirect Current Control of a Unity Power Factor Sinusoidal Current Boost Type Three-phase Rectifier, *IEEE Transaction of Indisturial Electronics,* Vol.35, Nov, 1988, pp. 508-515, ISSN 0278-0046

Dixon, J,W. (1990). Feedback Control Strategies for Boost Type PWM Rectifiers . *Proceedings of IEEE Colloquium in South America,* pp. 193-198, ISBN 0-87942-610-1, September, 1990

Fitzgerald, A,E.; Kingsley, C,Jr. & Umans, S,D. (2003). *Electric Machinery,* McGraw Hill, ISBN 0-07-112193-5, New York, USA

Holtz, J. (1994). Pulse Width Modulation for Electronic Power Conversion, *Proceedings of the IEEE,* Vol 82, August 1994, pp. 1194-1214, ISSN 0018-9219

Isidori, A. (1995). *Nonlinear Control Systems,* Springer-Verlag, ISBN 3-540-19916-0, Heidelberg New York

Kazmierkowski, M,P.; Krishnan, R. & Blaabjerg, F. (2002). *Control in Power Electronics: Selected Problems,* Elsevier Science, ISBN 0-12-402772-5, San Diego, California, USA

Kim, D,E. & Lee, D,C. (2007). Feedback Linearization Control of Three-Phase Ac/Dc PWM Converters with LCL Input Filters, *International Conference on Power Electronics ICPE'07*, pp. 766-771, ISBN 978-1-4244-1871-8, Daegu, South Korea, October, 2007

Khalil, H,K. (2000). *Nonlinear Systems*, Pearson Education, ISBN 0-13-122740-8, New Jersey, USA

Krishnan, R. (2001). *Electric Motor Drives: Modeling, Analysis, and Control*, Prentice Hall, ISBN 0-13-091014-7, New Jersey, USA

Kömürcügil, H. & Kükrer, O. (1998). Lyapunov Based Control of Three-Phase PWM Ac-Dc Voltage-Source Converters, *IEEE Transactions on Power Electronics*, Vol 13, September 1998, pp. 801-813, ISSN 0885-8993

Lee, D,C.; Lee, G,M. & Lee, K,D. (2000). Dc Bus Voltage Control of Three Phase Ac-Dc PWM Converters Using feedback Linearization, *IEEE Transactions on Industry Application*, Vol 36, May-June 2000, pp. 826-833, ISSN 0993-9994

Lee, T,S. (2003). Input-Output Linearizing and Zero Dynamics Control of Three-Phase Ac-Dc Voltage Source Converters, *IEEE Transactions on Power Electronics*, Vol 18, January 2003, pp. 11-22, ISSN 0885-8993

Lindgren, M. (1998). Modelling and Control of Voltage Source Converters Connected to the Grid, *PHDs Thesis Faculty of Chalmers University of Technology*, ISBN 91-7197-710-4, Goteborg, SWEEDEN, November, 1998

Liserre, M.; Blaabjerg, F. & Hansen, S. (2005). Design and Control of an LCL Filter Based Three-Phase Active Rectifier, *IEEE Transactions on Industry Application*, Vol 41, September-October 2005, pp. 1281-1291, ISSN 0993-9994

Mao, H.; Boroyevich, D. & Lee, F.C. (1998). Novel Reduced-Order Small-Signal Model of Three-Phase PWM Rectifiers and its Application in Control Design and System Analysis, *IEEE Transactions on Power Electronics*, Vol 13, May 1998, pp. 511-521, ISSN 0885-8993

Mihailovic, Z. (1998). Modeling and Control Design of VSI_ Fed PMSM Drive Systems with Active Load, *Master Thesis Faculty of Virginia Polytechnic and State University*, Blacksburg, Virginia, USA, September, 1998

Ooi, B,T.; Salmon, J,C. ; Dixon, J,W. & Kulkarini, A,B. (1987). A three-phase Controlled Current PWM Converter with Leading Power Factor, *IEEE Transaction of Indistury Application*, Vol.IA-23, Jan-Feb 1987, pp. 78-81, ISSN 0093-9994

Sehirli, E. & Altınay, M. (2010). Simulation of Three-Phase Voltage Source Pulse Width Modulated (PWM) LCL Filtered Rectifier Based on Input-Output Linearization Nonlinear Control, *IEEE International Conference on Electrical and Electronic Equipment OPTIM'2010*, pp. 564-569, ISBN 978-1-4244-7019-8, Brasov, Romania, May, 2010

Slotine, J,J,E. & Li, W. (1991). *Applied Nonlinear Control*, Prentice Hall, ISBN 0-13-040890-5, New Jersey, USA

Sousa, G,C,D. & Bose, B. (1994). A Fuzzy Set Theory Based Control of a Phase - Controlled Converter Dc Machine Drive, *IEEE Transactions on Industry Application*, Vol 30, January-February, 1994, pp. 34-44, ISSN 0993-9994

Wu, R.; Dewan, S,B. & Slemon, G,R. (1988). A PWM ac to dc Converter with Fixed Switching Frequency, *Conference Recordings 1988 IEEE-IAS Annual Meeting*, Vol 1, October, 1988 pp. 706-711, Pittsburgh, PA, USA

Wu, R.; Dewan, S,B. & Slemon, G,R. (1991). Analysing of ac to dc Voltage Source Converter Using PWM with Phase and Amplitude Control, *IEEE Transactions Industirial Applications,* Vol 27, March-April, 1991, pp. 355-364, ISSN 0093-9994

Ye, Z. (2000). Modelling and Control of Parallel Three-Phase PWM Converters, *PHDs Thesis Faculty of Virginia Polytechnic and State University,* pp. 9-20, Blacksburg, Virginia, USA, September, 2000

2

Lyapunov-Based Robust and Nonlinear Control for Two-Stage Power Factor Correction Converter

Seigo Sasaki
National Defense Academy
Japan

1. Introduction

Many power electronic system designs focus on energy conversion circuit parameters rather than controller parameters which drive the circuits. Controllers must be designed on the basis of circuit models, which are generally nonlinear systems (Brockett & Wood (1974)), in order to improve performance of controlled systems. The performance of controlled systems depend on nominal models to design the controllers. More broad class of models controllers are designed for, better control performance may be given. Many works (e.g. Kassakian et al. (1991)) design controllers for linearized models because it is not easy to concretely design controllers for the nonlinear models. Controller design in consideration of nonlinear models has been discussed since a work by Banerjee & Verghese (2001) because a research on nonlinear controller design has grown in those times.

This chapter systematically designs a robust controlled power converter system on the basis of its nonlinear model. Concretely, a two-stage power factor correction converter, that is a forward converter (FC) with power factor corrector (PFC), is designed. The systematic controller design clearly analyzes the behavior of nonlinear system to improve the performance.

A work by Orabi & Ninomiya (2003) analyzes a stability of single-stage PFC for variations of controller gain on the basis of its nonlinear model. On the basis of the work by Orabi & Ninomiya (2003) that regards a load of PFC as constant, a work by Dranga et al. (2005) for a two-stage PFC focuses on a point that a load of PFC part is not pure resistive and analyzes a stability of the converter.

A work by Sasaki (2002) discusses an influence between a FC part and a PFC part in a two-stage PFC. A work by Sasaki (2009) clearly shows that a source current reference generator plays an important role in a synthesis for a single-stage PFC. For the works by Sasaki (2002; 2009), this chapter shows how to decide synthesis parameters of robust linear and nonlinear controllers for a two-stage PFC in more detail.

The controller synthesis step, that is this chapter, is organized as follows. First, the converter is divided into two parts which consists of a FC part and a PFC part by considering an equivalent circuit of transformer. The two parts depend on each other and are nonlinear systems. The FC part has an apparent input voltage which depends on an output voltage in the PFC part. On the other hand, the PFC part has an apparent load resistance which depends on an input current in the FC part. Second, the two parts of converter are treated as two independent converters by analyzing steady state in the converter and deciding a set point. Then, the above input voltage and load resistance are fixed on the set point. Controllers are designed for the two independent converters respectively. For the FC part which is a linear system at the second stage of converter, a robust linear controller is designed against variations of the apparent input voltage and a load resistance. For the PFC part which is a bilinear system, that is a class of nonlinear system, at the first stage, a robust nonlinear controller is designed against variations of the apparent load resistance that mean dynamic variations of the FC part at the second stage. Finally, computer simulations demonstrate efficiencies of the approach. It is also clarified that consideration of nominal load resistance for each part characterizes a performance of the designed robust controlled system.

2. Two-stage power factor correction converter

A controlled system for a two-stage power factor correction converter, that is a forward converter (FC) with power factor corrector (PFC), is systematically constructed as shown in Fig.1. The systematic controller design clearly analyzes the behavior of controlled converter system. In this section, first, an equivalent circuit of transformer divides the converter into two parts with FC and PFC. Averaged models for the two parts are derived respectively, which depend on each other. Second, steady state in the two parts is analyzed, which is used as a set point. Finally, each of parts is treated as an independent converter respectively.

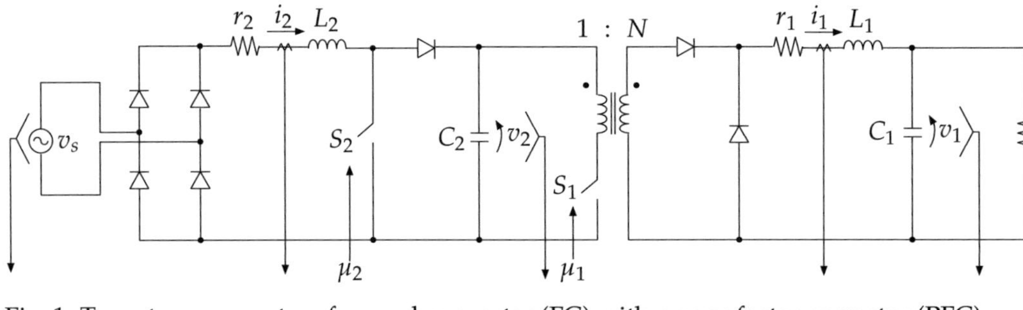

Fig. 1. Two-stage converter ; forward converter (FC) with power factor corrector (PFC)

2.1 Nonlinear averaged models

An equivalent circuit of transformer derives a circuit as shown in Fig. 2 from Fig. 1. A FC part has a variable dc source voltage which depends on a switch S_2. A PFC part has a variable load resistance which depends on a switch S_1. Fig. 2 gives two nonlinear averaged models (Σ_{SA}^{fc}) and (Σ_{SA}^{pfc}) for the FC and the PFC parts respectively, which depend on each other.

The FC model (Σ_{SA}^{fc}) is given as

$$\frac{d}{dt}\begin{bmatrix}\bar{v}_1\\ \bar{i}_1\end{bmatrix} = \begin{bmatrix}-\frac{1}{RC_1} & \frac{1}{C_1}\\ -\frac{1}{L_1} & -\frac{r_1}{L_1}\end{bmatrix}\begin{bmatrix}\bar{v}_1\\ \bar{i}_1\end{bmatrix} + \left\{\bar{v}_2\begin{bmatrix}0\\ \frac{N}{L_1}\end{bmatrix}\right\}\bar{\mu}_1 \tag{1}$$

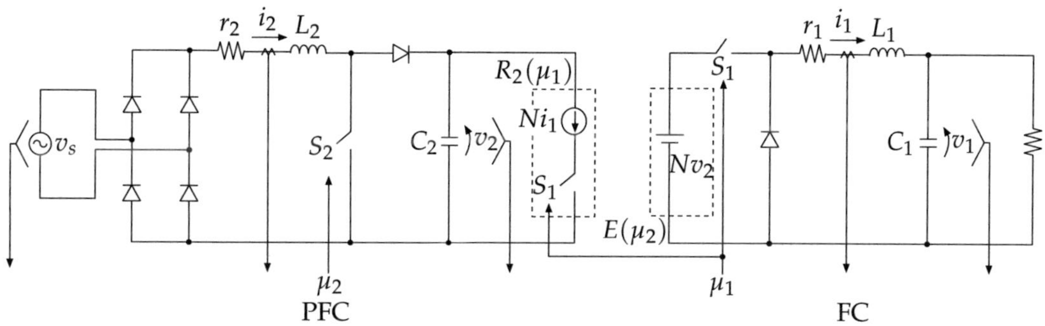

Fig. 2. Two converters ; FC and PFC

and the PFC model (Σ_{SA}^{pfc}) is

$$\frac{d}{dt}\begin{bmatrix}\bar{v}_2\\ \bar{i}_2\end{bmatrix} = \begin{bmatrix}0 & \frac{1}{C_2}\\ -\frac{1}{L_2} & -\frac{r_2}{L_2}\end{bmatrix}\begin{bmatrix}\bar{v}_2\\ \bar{i}_2\end{bmatrix} + \begin{bmatrix}0\\ \frac{1}{L_2}\end{bmatrix}|v_s|$$
$$+\left\{\bar{v}_2\begin{bmatrix}0\\ \frac{1}{L_2}\end{bmatrix} + \bar{i}_2\begin{bmatrix}-\frac{1}{C_2}\\ 0\end{bmatrix}\right\}\bar{\mu}_2 + \left\{\bar{i}_1\begin{bmatrix}-\frac{N}{C_2}\\ 0\end{bmatrix}\right\}\bar{\mu}_1 \tag{2}$$

where $\bar{v}_1$ and $\bar{v}_2$ are averaged capacitor voltages, $\bar{i}_1$,$\bar{i}_2$ averaged inductor currents, C_1,C_2 capacitances, L_1,L_2 inductances, r_1,r_2 internal resistances, R a load resistance, N a turns ratio, v_s a source voltage, $\bar{\mu}_1$,$\bar{\mu}_2$ averaged switching functions given by $0 < \bar{\mu}_1 < 1$ and $0 < \bar{\mu}_2 < 1$.

2.2 DC components of steady state

DC components of steady state in two parts of converter are derived, which are used as a set point for controller design. Given an average full-wave rectified voltage $V_s := (\omega/2\pi) \int_0^{\frac{2\pi}{\omega}} |v_s(t)|dt = 2\sqrt{2}V_e/\pi$ of a source voltage $v_s = \sqrt{2}\,V_e \sin\omega t$, and specify averaged switching functions $\bar{\mu}_1 = \bar{\mu}_{1s}$, $\bar{\mu}_2 = \bar{\mu}_{2s}$ which are called a set point, then dc components of voltages and currents of steady state are given by

$$\bar{v}_{1s} := \bar{\mu}_{1s}\frac{1}{1+\frac{r_1}{R}}E(\bar{\mu}_{2s}), \tag{3}$$

$$\bar{i}_{1s} := \frac{1}{R}\bar{v}_{1s} = \bar{\mu}_{1s}\frac{1}{1+\frac{r_1}{R}}\frac{E(\bar{\mu}_{2s})}{R}, \tag{4}$$

$$\bar{v}_{2s} := \frac{1}{1-\bar{\mu}_{2s}+\frac{1}{1-\bar{\mu}_{2s}}\frac{r_2}{R_2(\bar{\mu}_{1s})}}V_s, \tag{5}$$

$$\bar{i}_{2s} := \frac{1}{1-\bar{\mu}_{2s}}\bar{\mu}_{1s}N\bar{i}_{1s} = \frac{1}{1-\bar{\mu}_{2s}}\frac{\bar{v}_{2s}}{R_2(\bar{\mu}_{1s})} \tag{6}$$

where

$$E(\bar{\mu}_{2s}) := N\bar{v}_{2s}, \qquad R_2(\bar{\mu}_{1s}) := \frac{1+\frac{r_1}{R}}{\bar{\mu}_{1s}^2N^2}R. \tag{7}$$

The above steady states in two parts depend on each other through (7). It is easily clarified by assuming $r_1 = r_2 = 0$ that the equations (3), (4) give a steady state in buck converter and the equations (5), (6) give a steady state in boost converter.

2.3 Two independent models

For two parts of converter with FC and PFC, the steady state analysis derives two independent converters. It means that the FC has an apparent source voltage $E(\bar{\mu}_{2s})$ and the PFC has an apparent load resistance $R_2(\bar{\mu}_{1s})$ which are fixed on a set point respectively.

Then, two independent averaged models of the converters are given as

a linear FC model (Σ_{A0}^{fc}) of the form

$$\frac{d}{dt}\begin{bmatrix}\bar{v}_1\\ \bar{i}_1\end{bmatrix} = \begin{bmatrix}-\frac{1}{RC_1} & \frac{1}{C_1}\\ -\frac{1}{L_1} & -\frac{r_1}{L_1}\end{bmatrix}\begin{bmatrix}\bar{v}_1\\ \bar{i}_1\end{bmatrix} + \begin{bmatrix}0\\ \frac{E(\bar{\mu}_{2s})}{L_1}\end{bmatrix}\bar{\mu}_1 \tag{8}$$

and a nonlinear PFC model (Σ_{A0}^{pfc}) of the form

$$\frac{d}{dt}\begin{bmatrix}\bar{v}_2\\ \bar{i}_2\end{bmatrix} = \begin{bmatrix}-\frac{1}{R_2(\bar{\mu}_{1s})C_2} & \frac{1}{C_2}\\ -\frac{1}{L_2} & -\frac{r_2}{L_2}\end{bmatrix}\begin{bmatrix}\bar{v}_2\\ \bar{i}_2\end{bmatrix} + \begin{bmatrix}0\\ \frac{1}{L_2}\end{bmatrix}|v_s| + \left\{\bar{v}_2\begin{bmatrix}0\\ \frac{1}{L_2}\end{bmatrix} + \bar{i}_2\begin{bmatrix}-\frac{1}{C_2}\\ 0\end{bmatrix}\right\}\bar{\mu}_2. \tag{9}$$

The FC is modeled on a linear system and the PFC is modeled on a bilinear system that is a class of nonlinear system (Brockett & Wood (1974); Mohler (1991)).

Here derived is an amplitude of source current of steady state in the PFC, which is used for source current reference generator in Section 3.3. Given a source voltage $v_s = \sqrt{2}V_e \sin\omega t$ [V], assume an output voltage $\bar{v}_2 = v_{2r}$ [V], then dc components of steady state in the PFC gives a source current $\bar{i}_2 = \sqrt{2}I_e \sin\omega t$ [A], which is sinusoidal and in phase with the source voltage, given by

$$I_e = \frac{V_e}{2r_2} - \sqrt{\frac{V_e^2}{4r_2^2} - \frac{v_{2r}^2}{r_2 R_2(\bar{\mu}_{1s})}}. \tag{10}$$

This decision process is similar to that is discussed in works by Escobar et al. (1999); Escobar et al. (2001).

3. Robust controller design

Controllers for two parts with FC and PFC are designed on the basis of two independent converter models (Σ_{A0}^{fc}) and (Σ_{A0}^{pfc}) respectively. First, the models (Σ_{A0}^{fc}) and (Σ_{A0}^{pfc}) are moved to set points given in Section 2.2. Next, controller design models, that are called as generalized plants, are derived which incorporate weighting functions in consideration of controller design specifications. Finally, feedback gains are given to guarantee closed-loop stability and reference tracking performance.

3.1 Models around set points

Converter models around set points are derived from the averaged models (Σ_{A0}^{fc}) and (Σ_{A0}^{pfc}). Moving the state to a specified set point given by $\tilde{v}_1 := \bar{v}_1 - \bar{v}_{1s}, \tilde{i}_1 := \bar{i}_1 - \bar{i}_{1s}, \tilde{\mu}_1 := \bar{\mu}_1 - \bar{\mu}_{1s}$, $\tilde{v}_2 := \bar{v}_2 - \bar{v}_{2s}, \tilde{i}_2 := \bar{i}_2 - \bar{i}_{2s}, \tilde{\mu}_2 := \bar{\mu}_2 - \bar{\mu}_{2s}$ derives averaged models around the set point respectively for the two parts, which are given by

a FC model (Σ_A^{fc}) of the form

$$\frac{d}{dt}\begin{bmatrix}\tilde{v}_1\\ \tilde{i}_1\end{bmatrix} = \begin{bmatrix}-\frac{1}{RC_1} & \frac{1}{C_1}\\ -\frac{1}{L_1} & -\frac{r_1}{L_1}\end{bmatrix}\begin{bmatrix}\tilde{v}_1\\ \tilde{i}_1\end{bmatrix} + \begin{bmatrix}0\\ \frac{E(\bar{\mu}_{2s})}{L_1}\end{bmatrix}\tilde{\mu}_1 \tag{11}$$

$$=: A_p^{fc}x_p^{fc} + B_p^{fc}u_1 \tag{12}$$

and a PFC model (Σ_A^{pfc}) of the form

$$\frac{d}{dt}\begin{bmatrix}\tilde{v}_2\\ \tilde{i}_2\end{bmatrix} = \begin{bmatrix}-\frac{1}{R_2(\bar{\mu}_{1s})C_2} & \frac{1}{C_2}(1-\bar{\mu}_{2s})\\ -\frac{1}{L_2}(1-\bar{\mu}_{2s}) & -\frac{r_2}{L_2}\end{bmatrix}\begin{bmatrix}\tilde{v}_2\\ \tilde{i}_2\end{bmatrix} + \begin{bmatrix}0\\ \frac{1}{L_2}\end{bmatrix}(|v_s| - Vs)$$
$$+\left\{\begin{bmatrix}-\frac{\bar{i}_{2s}}{C_2}\\ \frac{\bar{v}_{2s}}{L_2}\end{bmatrix} + \tilde{v}_2\begin{bmatrix}0\\ \frac{1}{L_2}\end{bmatrix} + \tilde{i}_2\begin{bmatrix}-\frac{1}{C_2}\\ 0\end{bmatrix}\right\}\tilde{\mu}_2 \tag{13}$$

$$=: A_p^{pfc}x_p^{pfc} + B_{p1}^{pfc}w_p^{pfc} + \left(B_p^{pfc} + \left\{x_p^{pfc}N_p^{pfc}\right\}\right)u_2 \tag{14}$$

where $-\bar{\mu}_{1s} < \tilde{\mu}_1 < 1 - \bar{\mu}_{1s}$ and $-\bar{\mu}_{2s} < \tilde{\mu}_2 < 1 - \bar{\mu}_{2s}$. The following discussion constructs controllers on the basis of the averaged models (Σ_A^{fc}) and (Σ_A^{pfc}) around the set point.

3.2 Linear controller design for FC

A robust controller for the FC is given by linear H^∞ control technique because the averaged model (Σ_A^{fc}) is a linear system. The control technique incorporates weighting functions $W_{v1}(s) := k_{v1}/(s+\varepsilon_1)$ to keep an output voltage constant against variations of load resistance and apparent input voltage as shown in Fig. 3, where k_{v1}, ε_1 are synthesis parameters. Then, the linear H^∞ control technique gives a linear gain K_1 of the form

$$u_1 = K_1x^{fc} := -(D_{12}^{fc^T}D_{12}^{fc})^{-1}B_2^{fc^T}Y^{fc^{-1}}x^{fc} \tag{15}$$

where $x^{fc} := \begin{bmatrix}x_p^{fc^T} & x_w^{fc^T}\end{bmatrix}^T$, $B_2^{fc} := \begin{bmatrix}B_p^{fc^T} & 0^T\end{bmatrix}^T$, $D_{12}^{fc} := \begin{bmatrix}0^T & W_{u1}^T\end{bmatrix}^T$ and x_w^{fc} denotes state of weighting function $W_{v1}(s)$ as shown in Fig.3. In the figure, matrices W_{e1} and W_{u1} are weighting coefficients of a performance index in the linear H^∞ control technique and used only in the controller synthesis.

The matrix Y^{fc} that constructs the gain K_1 is given as a positive-definite solution satisfying a Lyapunov-based inequality condition

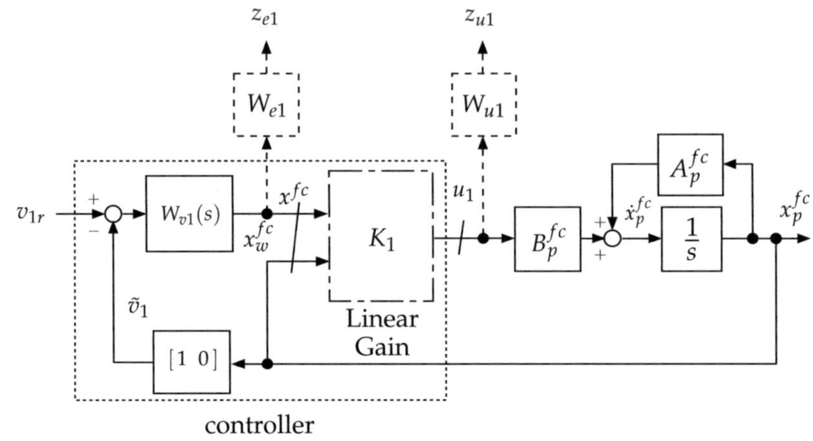

Fig. 3. Block diagram of linear controlled system for FC

$$\begin{bmatrix} A^{fc}Y^{fc} + Y^{fc}A^{fc^T} - B_2^{fc}(D_{12}^{fc^T} D_{12}^{fc})^{-1} B_2^{fc^T} + \gamma^{fc^{-2}} B_1^{fc} B_1^{fc^T} & Y^{fc}C_1^{fc^T} \\ C_1^{fc}Y^{fc} & -I \end{bmatrix} < 0 \tag{16}$$

where the matrix A^{fc},B_1^{fc}, C_1^{fc} are coefficients of a generalized plant given by Fig.3 and γ^{fc} is a synthesis parameter in the linear H^∞ control technique.

3.3 Nonlinear controller design for PFC

A robust nonlinear controller design for the PFC, that is a nonlinear system, consists of the following two steps. First, a nonlinear gain is given to guarantee closed loop system stability and reference tracking performance for source current and output voltage against variations of the apparent load resistance. Second, a source current reference generator is derived to adjust an amplitude of current reference to variations of source voltage and load resistance. At the first step, as shown in Fig. 4 incorporated are a weighting function $W_{i2}(s) := k_{i2}\omega_{i2}^2/(s^2 + 2\zeta\omega_{i2}s + \omega_{i2}^2)$ for a source current to be sinusoidal and be in phase with a source voltage and a function $W_{v2}(s) := k_{v2}/(s + \varepsilon_2)$ for an output voltage to be kept constant against those variations, where k_{i2}, ζ, ω_{i2}, k_{v2}, ε_2 are synthesis parameters. Then, a nonlinear H^∞ control technique in a work by Sasaki & Uchida (1998) gives a nonlinear gain of the form

$$K_2(x_p^{pfc}) := -(D_{12}^{pfc^T} D_{12}^{pfc})^{-1} B_2^{pfc}(x^{pfc})^T Y^{pfc^{-1}} \tag{17}$$

as shown in Fig.4, where $x^{pfc} = \begin{bmatrix} x_p^{pfc^T} & x_w^{pfc^T} \end{bmatrix}^T$, $D_{12}^{pfc} = \begin{bmatrix} 0^T & W_{u2}^T \end{bmatrix}^T$, $B_2^{pfc}(x^{pfc}) = \begin{bmatrix} B_p^{pfc} \\ 0 \end{bmatrix} + x_1^{pfc} \begin{bmatrix} B_{p21}^{pfc} \\ 0 \end{bmatrix} + x_2^{pfc} \begin{bmatrix} B_{p22}^{pfc} \\ 0 \end{bmatrix}$, $B_{p21}^{pfc} = \begin{bmatrix} 0 & \frac{1}{L_2} \end{bmatrix}^T$, $B_{p22}^{pfc} = \begin{bmatrix} -\frac{1}{C_2} & 0 \end{bmatrix}^T$ and x_w^{pfc} denotes state of weighting functions. In the figure, matrices W_{e2} and W_{u2} are weighting coefficients of a performance index in the nonlinear H^∞ control technique. A block named as Generator is not used for the synthesis and is discussed at the following second step.

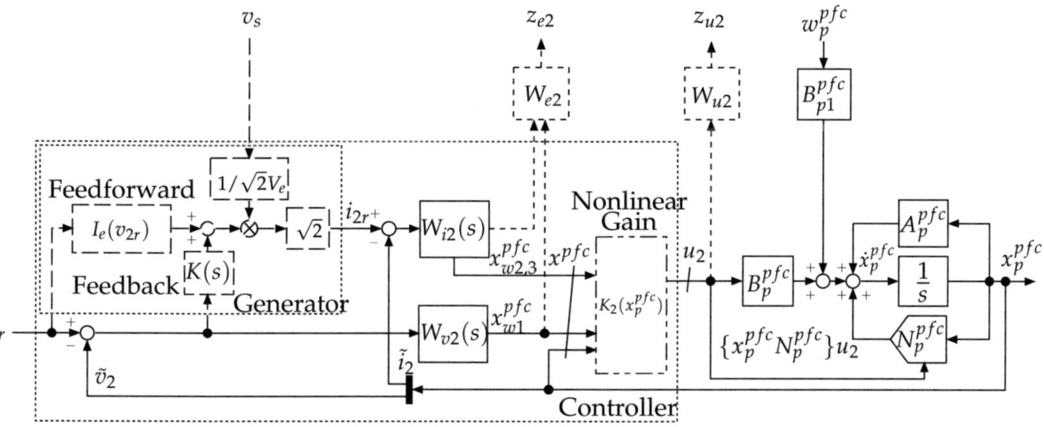

Fig. 4. Block diagram of nonlinear controlled system for PFC

The matrix Y^{pfc} is given as a positive-definite solution satisfying a state-depended Lyapunov-based inequality condition

$$\begin{bmatrix} A^{pfc}Y^{pfc} + Y^{pfc}A^{pfc^T} - B_2^{pfc}(x)(D_{12}^{pfc^T}D_{12}^{pfc})^{-1}B_2^{pfc}(x)^T & Y^{pfc}C_1^{pfc^T} \\ +\gamma^{pfc^{-2}}B_1^{pfc}B_1^{pfc^T} & \\ C_1^{pfc}Y^{pfc} & -I \end{bmatrix} < 0 \qquad (18)$$

where x is a state of a generalized plant given by Fig.4, matrices A^{pfc},B_1^{pfc},$B_2^{pfc}(x)$,C_1^{pfc},D_{12}^{pfc} are coefficients of the plant and γ^{pfc} is a synthesis parameter in the nonlinear H^∞ control technique.

For any state x, which is current and voltage in a specified domain, the matrix Y^{pfc} satisfying the inequality (18) is concretely given by solving linear matrix inequalities at vertices of a convex hull enclosing the domain as shown in a work by Sasaki & Uchida (1998).

Next, given is a mechanism to generate a source current reference i_{2r}. As shown in Fig.4 the reference generator consists of a feedforward loop given by steady state analysis in Section 2.3 and a feedback loop with a voltage error amplifier. The amplifier $K(s)$ is given by $K(s) = k_P + k_I/s + k_D s$ where k_P,k_I and k_D are constant parameters decided by system designers as discussed in a work by Sasaki (2009). The feedback loop is the same structure as a conventional loop used in many works (e.g., Redl (1994)). Note that the amplifier $K(s)$ works only for variations from the nominal values in the circuit, because the feedforward loop gives the effective value of the source current as shown in Section 4.4.

4. Computer simulations

This section finally shows efficiencies of the approach through computer simulations. It is also clarified that consideration of nominal load resistance for each part characterizes a performance of the designed controlled system. A software package that consists of MATLAB, Simulink and LMI Control Toolbox is used for the simulations.

Parameters of the circuit shown in Fig.1 are given by Table 1.

r_1 1 [mΩ]	r_2 1 [mΩ]
L_1 10 [mH]	L_2 1 [mH]
C_1 450 [μF]	C_2 1000 [μF]
N 1/36	

Table 1. Circuit parameters of two-stage power factor correction converter as shown in Fig.1

4.1 Design specification

Control system design specification for the two-stage power factor correction converter is given by

(1) An effective value of source voltage is 82 $\sim$ 255 volts, and its frequency is 45 $\sim$ 65 hertz ;

(2) An output voltage is kept be 5 volts, and its error of steady state is within $\pm 0.5\%$;

(3) An output current is 0 $\sim$ 30 amperes ;

(4) A source current is approximately sinusoidal and is phase with source voltage.

The above specifications are treated for controller design as the following respective considerations ;

(1) A nonlinear gain for PFC is designed for a source voltage whose nominal effective value is $V_e = (82 + 255)/2 = 168.5$ volts (then, average full-wave rectified voltage is $V_s = 151.7$ volts). Moreover, the gain is designed to be robust against source voltage variations by treating that as a disturbance w_p^{pfc} as shown in Fig.4.

(2) Incorporated is a weighting function $W_{v1}(s)$ whose magnitude is high at low frequencies.

(3) A load resistances R varies between $1/6 \sim \infty$ ohms at an output voltage of 5 volts.

(4) Incorporated is a weighting function $W_{i2}(s)$ whose magnitude is high at frequencies in the range of 45 $\sim$ 65 hertz for the source current to be approximately sinusoidal at the frequencies.

4.2 Nominal load resistance

Now, a nominal value of load resistance R need be decided to design controllers. The nominal value influences a performance of closed-loop system controlled by the designed controllers. Root loci of linearized converters for the two parts as shown in Figs. 5 and 6 , which are pointwise eigenvalues of the system matrices for variations of load resistance R, show that the larger the load resistance R is, the more oscillatory the behavior of FC is and the slower the transient response in PFC is. Controllers, here, are constructed in consideration of undesirable behavior of the converter system. Therefore, a FC controller is designed for a system with oscillatory behavior, that is for a large load resistance, so that an output voltage does not oscillate. A PFC controller is designed for a system with fast response, that is for a small load resistance, so that a source current is not distorted by the controller responding well to source voltage variations.

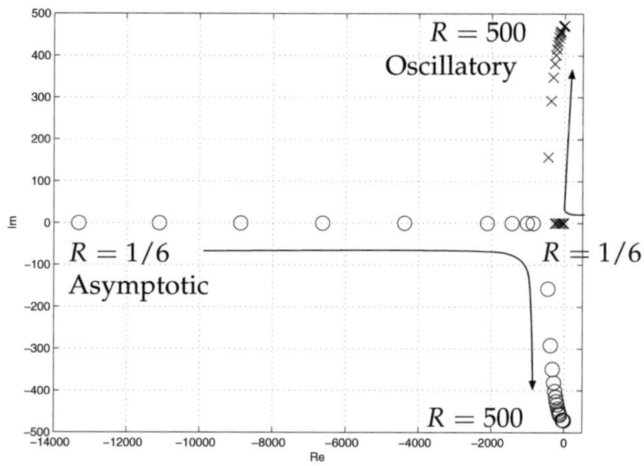

Fig. 5. Root locus of coefficient A_p^{fc} of linear model for FC

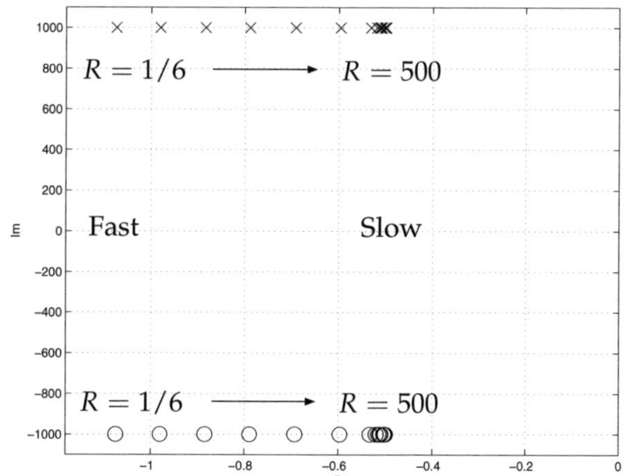

Fig. 6. Root locus of coefficient A_p^{pfc} of linearized model for PFC

4.3 Linear controller design parameters for FC

A FC controller is designed for a large nominal load resistance $R = 100$ ohms. Consider a set point $\bar{\mu}_{1s} = 0.5$ for a capacitor voltage of $\bar{v}_{2s} = 360$ volts. DC components of steady state are given by $\bar{v}_{1s} = 4.99$ volts and $\bar{i}_{1s} = 0.0499$ amperes. For a model around the above set point, incorporated are weighting functions with parameters given by $K_{v1} = 20, \varepsilon_1 = 0.001$ $W_{e1} = 5$, $W_{u1} = 1.5$ and $\gamma^{fc} = 0.95$ where γ^{fc} denotes control performance level given in the linear H^∞ control technique. The weighting function $W_{v1}(s)$ with the above parameter gives bode plots whose characteristics is like an integrator as shown in Fig.7.

Then, the matrix Y^{fc} that gives a linear gain (15) is obtained as

$$Y^{fc} = \begin{bmatrix} 2.00 \times 10^4 & -1.55 \times 10^3 & 5.05 \times 10^2 \\ -1.55 \times 10^3 & 7.68 \times 10^2 & 2.59 \times 10^1 \\ 5.05 \times 10^2 & 2.59 \times 10^1 & 2.51 \times 10^1 \end{bmatrix}. \tag{19}$$

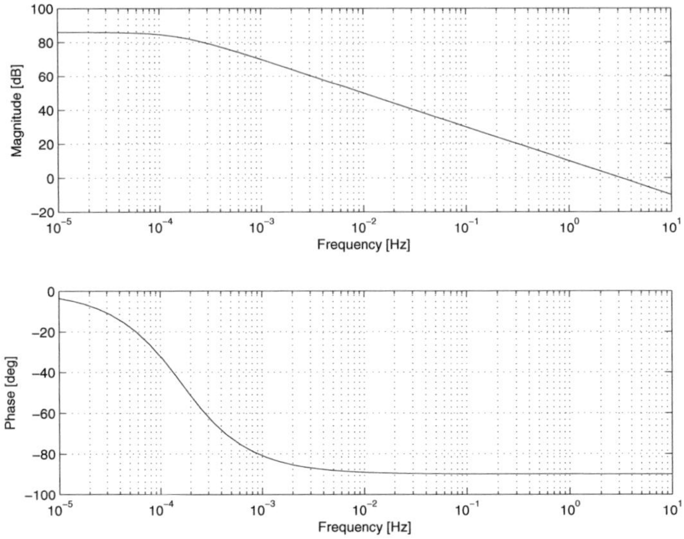

Fig. 7. Bode plots of weighting function $W_{v1}(s)$ for output voltage tracking in FC controller design

4.4 Nonlinear controller design parameters for PFC

A PFC controller is designed for a small nominal load resistance $R = 10$ ohms. Consider a set point $\bar{\mu}_{2s} = (v_{2r} - V_s)/v_{2r} = 0.579$ to make a rectified voltage $v_{2r} = 360$ volts. Then, an apparent load resistance in the PFC is given by $R_2(\bar{\mu}_{1s}) = 51845.2$ ohms for a set point $\bar{\mu}_{1s} = 0.5$ in the FC. Therefore, dc components of steady state are given by $\bar{v}_{2s} = 359.9$ volts and $\bar{i}_{2s} =$

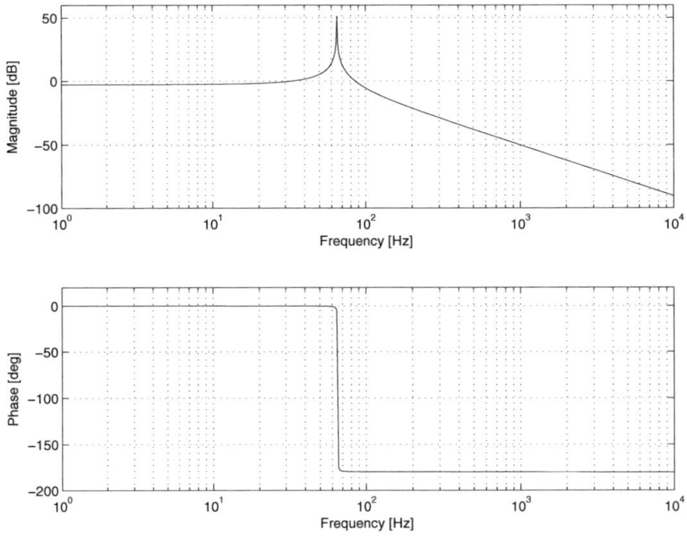

Fig. 8. Bode plots of weighting function $W_{i2}(s)$ for achieving unity power factor in PFC controller design

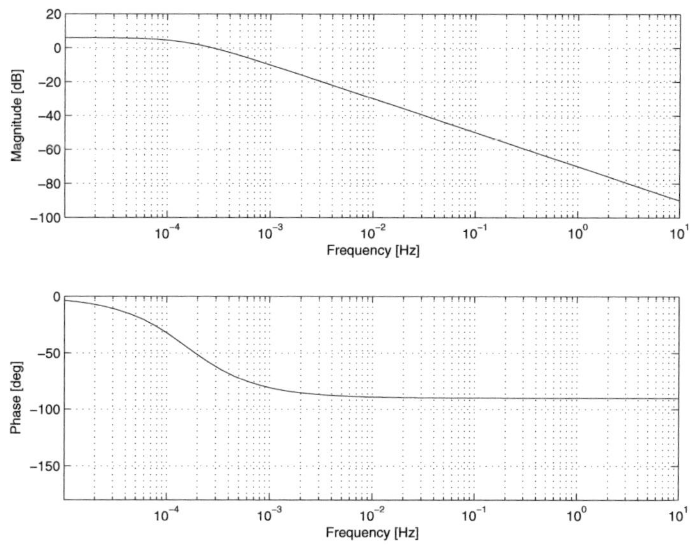

Fig. 9. Bode plots of weighting function $W_{v2}(s)$ for capacitor voltage regulation in PFC controller design

0.0165 amperes. For a model around the above set point, incorporated are weighting functions with parameters given by $\zeta = 0.001$, $\omega_{i2} = 130\pi$, $K_{i2} = 300/\omega_{i2}$, $K_{v2} = 0.002$, $\varepsilon_2 = 0.001$, $W_{e2} = \mathrm{diag}\,[10^{-7},\ 1]$, $W_{u2} = 20$ and $\gamma^{pfc} = 0.98$ where γ^{pfc} denotes control performance level given in the nonlinear H^{∞} control technique. The weighting function $W_{i2}(s)$ with the above parameter gives bode plots that is weighted at frequencies given by the design specification (4). The function $W_{v2}(s)$ has low gains at all frequencies because a voltage tracking fed-back by a controller is not important for the PFC design. It also leads that the W_{e2} is set such that the tracking performance for source current is more weighted than that for output voltage.

Then the matrix Y^{pfc} that gives a nonlinear gain (17) is obtained as

$$Y^{pfc} = \begin{bmatrix} 2.77\times10^{4} & -5.56\times10^{4} & 2.93\times10^{0} & 4.70\times10^{3} & -1.72\times10^{4} \\ -5.56\times10^{4} & 1.89\times10^{5} & 8.86\times10^{-1} & -5.57\times10^{3} & 4.03\times10^{4} \\ 2.93\times10^{0} & 8.86\times10^{-1} & 9.71\times10^{8} & -2.60\times10^{-2} & 1.89\times10^{-1} \\ 4.70\times10^{3} & -5.57\times10^{3} & -2.60\times10^{-2} & 1.10\times10^{3} & -2.98\times10^{3} \\ -1.72\times10^{4} & 4.03\times10^{4} & 1.89\times10^{-1} & -2.98\times10^{3} & 1.24\times10^{4} \end{bmatrix} \quad (20)$$

by considering a voltage variation of ±10 volts and a current variation of ±3 amperes around the set point (i.e., by considering $349.9 \le v_2 \le 369.9$ and $-2.9835 \le i_2 \le 3.0165$) and solving a convex programming problem as shown in the work by Sasaki & Uchida (1998).

Next, a source current reference generator is constructed with a voltage error amplifier given by $k_P = 8\times10^{-2}, k_I = 4, k_D = 1\times10^{-5}$ as shown in Fig.10. The gain is chosen in order to be low around twice the input line frequency.

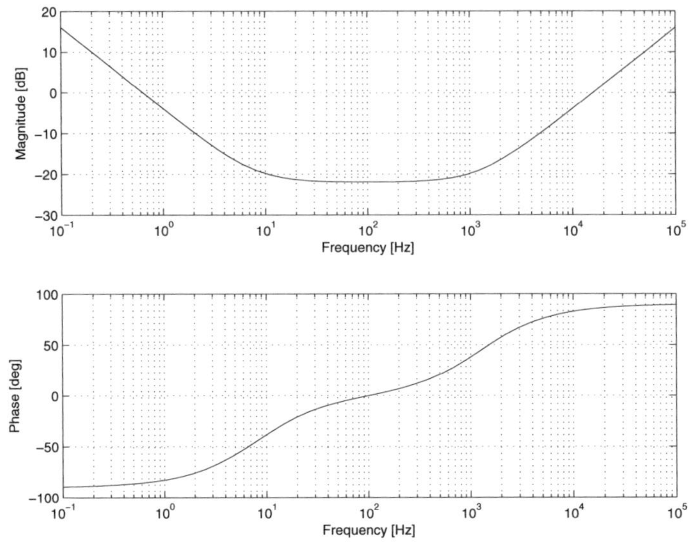

Fig. 10. Bode plots of voltage error amplifier $K(s)$ in PFC controller

4.5 Simulation results

Figs. 11–13 show behaviors of the nonlinear averaged models of FC (Σ_{SA}^{fc}) and PFC (Σ_{SA}^{pfc}) in the following cases ;

(C1)A load resistance R changes from 1000 to 0.25 ohms in steady state for a source voltage $v_s = \sqrt{2}\,100 \sin 100\pi t$ volts ;

(C2)An efficient value V_e of source voltage with 50 hertz changes from 100 to 85 volts for 2 seconds in steady state with a load resistance $R = 0.25$ ohms.

Figs. 11–13 show that the controllers works very well.

Fig.12 shows behaviors by a different controller from that in Figs.11 and 13, which is designed for desirable load resistances with a small nominal resistance $R = 10$ ohms for the FC and a large $R = 100$ ohms for the PFC.

Therefore, in the FC the output voltage v_1 in Fig. 12 oscillates more than in Fig. 11 at the large resistance $R = 1000$. Also, in the PFC the capacitor voltage v_2 in Fig. 12 drops larger than in Fig. 11. Then the source current i_2 in Fig.12 raises faster than in Fig. 11 and then is more distorted.

In the case as shown in Fig. 12, the above control techniques give the following matrices as

$$Y^{fc} = \begin{bmatrix} 2.92 \times 10^4 & -1.28 \times 10^3 & 6.26 \times 10^2 \\ -1.28 \times 10^3 & 8.07 \times 10^2 & 3.57 \times 10^1 \\ 6.26 \times 10^2 & 3.57 \times 10^1 & 2.27 \times 10^1 \end{bmatrix} \tag{21}$$

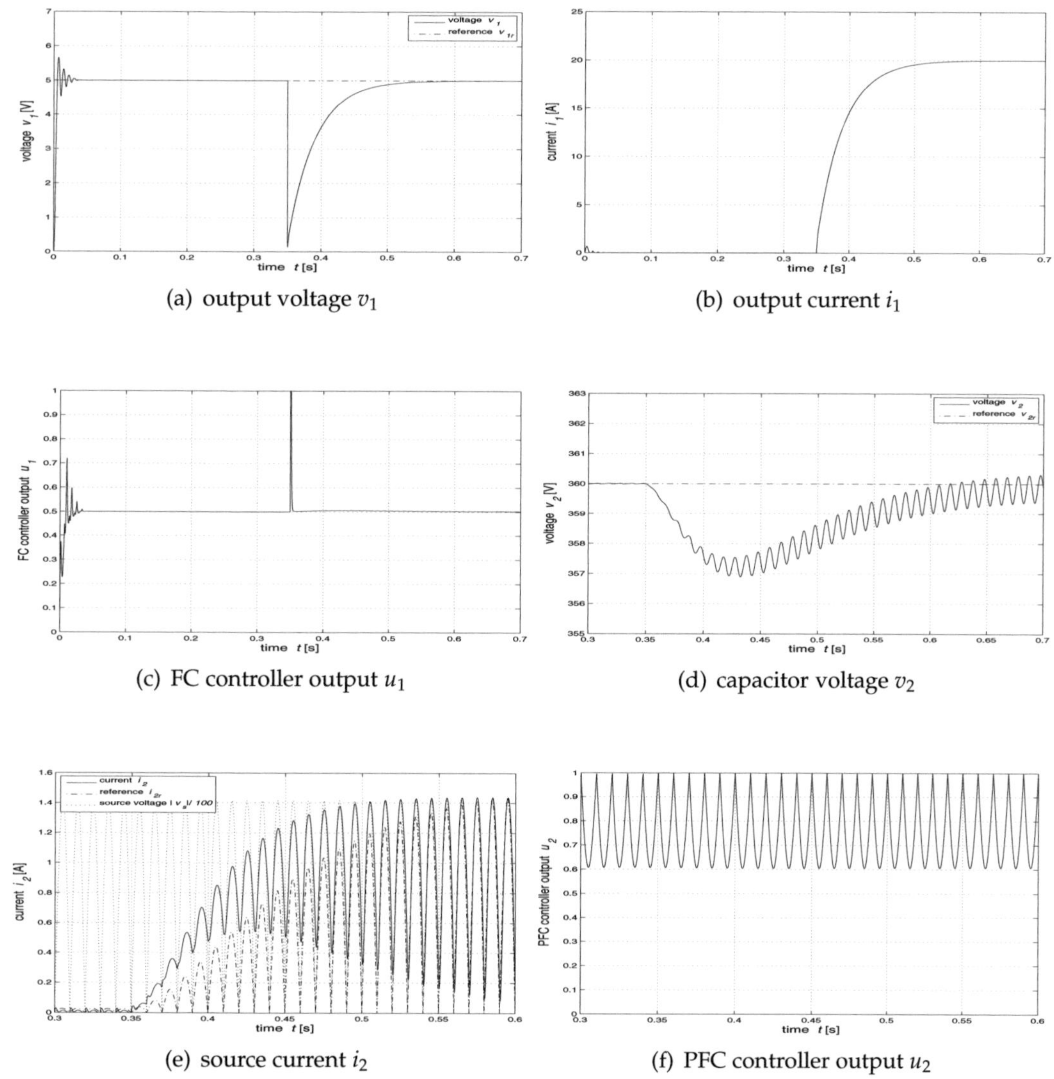

(a) output voltage v_1

(b) output current i_1

(c) FC controller output u_1

(d) capacitor voltage v_2

(e) source current i_2

(f) PFC controller output u_2

Fig. 11. (C1) Behaviors when a load resistance R changes from 1000 to 0.25 ohms in steady state

and

$$Y^{pfc} = \begin{bmatrix} 2.76 \times 10^4 & -5.56 \times 10^4 & 2.93 \times 10^0 & 4.70 \times 10^3 & -1.72 \times 10^4 \\ -5.56 \times 10^4 & 1.89 \times 10^5 & 8.85 \times 10^{-1} & -5.56 \times 10^3 & 4.03 \times 10^4 \\ 2.93 \times 10^0 & 8.85 \times 10^{-1} & 9.71 \times 10^8 & -2.60 \times 10^{-2} & 1.89 \times 10^{-1} \\ 4.70 \times 10^3 & -5.56 \times 10^3 & -2.60 \times 10^{-2} & 1.10 \times 10^3 & -2.98 \times 10^3 \\ -1.72 \times 10^4 & 4.03 \times 10^4 & 1.89 \times 10^{-1} & -2.98 \times 10^3 & 1.24 \times 10^4 \end{bmatrix}. \quad (22)$$

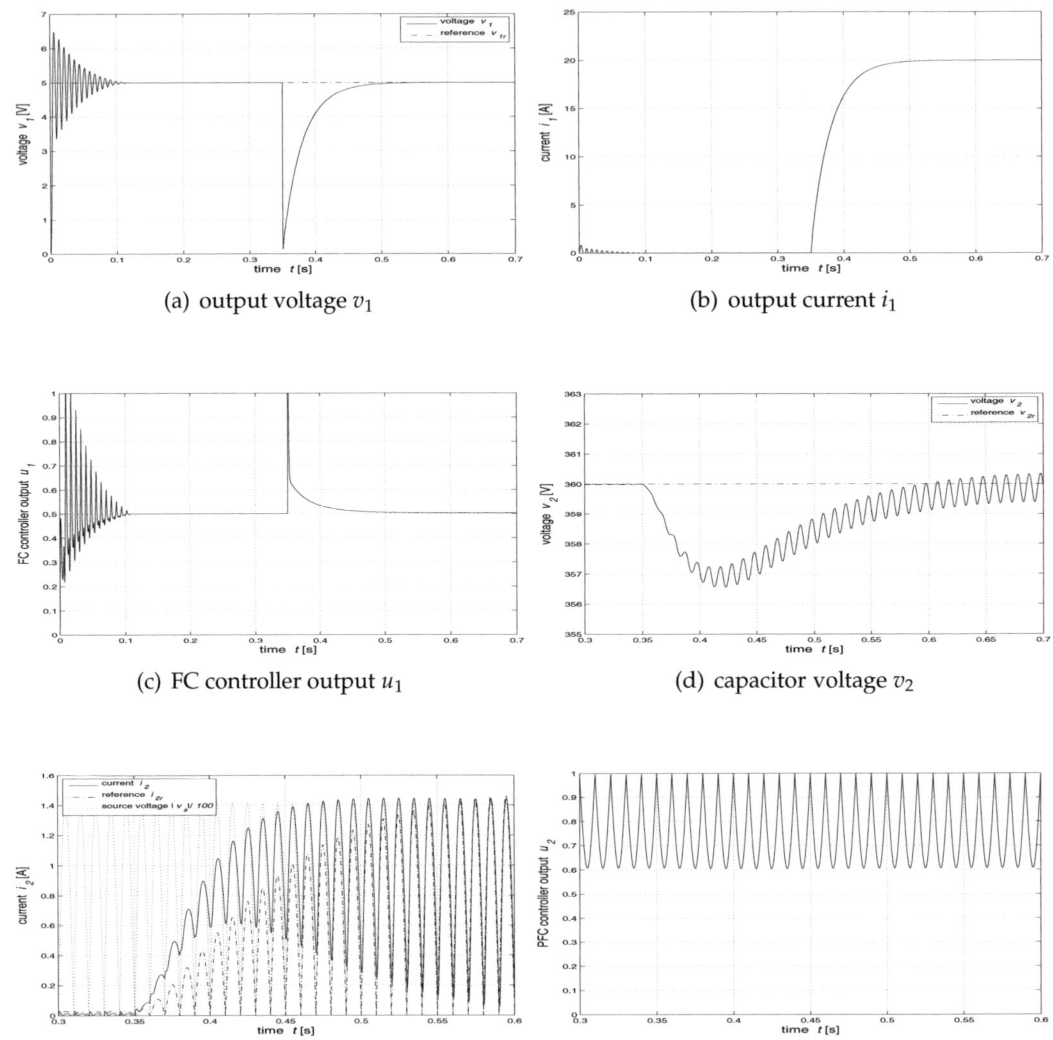

(a) output voltage v_1 (b) output current i_1

(c) FC controller output u_1 (d) capacitor voltage v_2

(e) source current i_2 (f) PFC controller output u_2

Fig. 12. (C1) Behaviors when a load resistance R changes from 1000 to 0.25 ohms in steady state where a controller is designed for a different nominal load resistance from that in Fig.11

Entries of the matrix Y^{fc} in (21) are clearly different from those in (19). On the other hand, difference of the Y^{pfc} between the matrix (22) and (20) is much small. Those differences give the different behaviors as mentioned above.

Fig.12 demonstrates the discussion in Section 4.2 that controllers needs be designed in consideration of undesirable behavior of system. For a large load resistance the output voltage of the FC oscillates and for small load resistance the source current and the capacitor voltage of the PFC respond to variations larger than those in Fig.11.

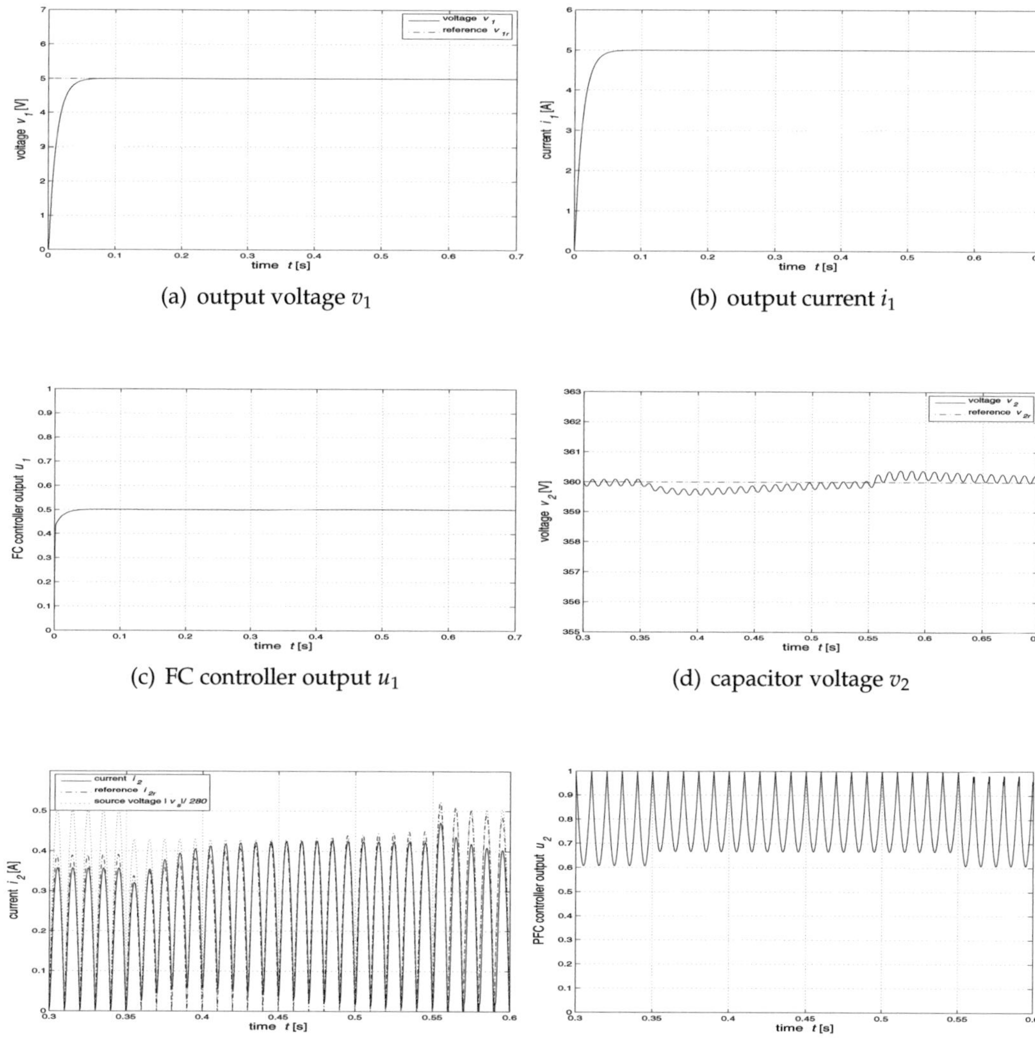

(a) output voltage v_1

(b) output current i_1

(c) FC controller output u_1

(d) capacitor voltage v_2

(e) source current i_2

(f) PFC controller output u_2

Fig. 13. (C2) Behaviors when an efficient value of source voltage V_e changes from 100 to 85 volts for 2 seconds in steady state

5. Conclusion

For a two-stage power factor correction converter, a nonlinear controlled converter system was systematically designed on the basis of its nonlinear model. The systematic controller design clearly analyzed the behavior of nonlinear system to improve the performance. The synthesis step were clearly shown. It was clarified that a nominal load resistance characterizes the controller performance. Finally, computer simulations demonstrated efficiencies of the approach. The nonlinear controller synthesis was shown as a natural extension of a well-known Lyapunov-based linear controller synthesis.

6. Acknowledgment

The author would like to thank Hideho Yamamura of Hitachi,Ltd. and Kazuhiko Masamoto of NEC Networks for their valuable discussions and comments in the viewpoint of industrial practice.

7. References

Brockett, R.W. & Wood, J.R. (1974). Electrical Networks Containing Controlled Switches. *IEEE Symposium on Circuit Theory*, pp.1-11

Kassakian, J.G.; Schlecht, M.F. & Verghese, G.C. (1991). *Principles of Power Electronics*, Addison-Wesley, ISBN:0-201-09689-7

Banerjee, S. & Verghese, G.C. (2001). *Nonlinear Phenomena in Power Electronics*, IEEE Press, Piscataway, NJ, ISBN:0-7803-5383-8

Orabi, M. & Ninomiya, T. (2003). Nonlinear Dynamics of Power-Factor-Correction Converter. *IEEE Transactions on Industrial Electronics*, pp.1116-1125

Dranga, O.;Tse, C.K. & Siu Chung Wong (2005). Stability Analysis of Complete Two-Stage Power-Factor-Correction Power Supplies *Proc. European Conference on Circuit Theory and Design*, pp.I/177-I/180

Mohler, R.R. (1991). *Nonlinear Systems, Vol.2, Applications to Bilinear Control*. Prentice Hall, Englewood Cliffs, NJ, ISBN:0-13-623521-2

Escobar, G.; Chevreau, D.; Ortega, R. & Mendes, E. (1999). An Adaptive Passivity-Based Controller for a Power Factor Precompensator. *Proc. 5th European Control Conference*, BP4-1

Escobar, G.; Chevreau, D.; Ortega,R. & Mendes, E. (2001). An Adaptive Passivity-Based Controller for a Unity Power Factor Rectifier. *IEEE Trans. Control Systems Technology*, Vol.9, No.4, pp.637-644

Sasaki, S. & Uchida, K. (1998). Nonlinear H_∞ Control System Design via Extended Quadratic Lyapunov Function. *Proc. IFAC Nonlinear Control Systems Design Symposium*, pp.163-168

Sasaki, S. (2002). System Clarification through Systematic Controller Design for a Forward Converter with Power Factor Corrector. *IEEE 33rd Annual Conf. on Power Electronics Specialists Conference*, Vol.3, pp.1083-1088

Sasaki,S. (2009). Systematic Nonlinear Control Approach to a Power Factor Corrector Design. *European Transactions on Electrical Power*, Vol.19, No.3, pp.460-473

Redl, R. (1994). Power-Factor Correction in Single-Phase Switching-Mode Power Supplies – An Overview. *International Journal of Electronics*, pp.555-582

Nonlinear Control Applied to the Rheology of Drops in Elongational Flows with Vorticity

Israel Y. Rosas[1], Marco A. H. Reyes[2], A. A. Minzoni[3] and E. Geffroy[1]
[1]*Instituto de Investigaciones en Materiales,*
[2]*Facultad de Ingeniería,*
[3]*Instituto de Investigaciones en Matemáticas Aplicadas y Sistemas*
Universidad Nacional Autónoma de México, Mexico City
Mexico

1. Introduction

Fluid systems where a liquid phase is dispersed in other liquid, as emulsions, are present in many industrial processes, technological applications and in natural systems. The flow of these substances shows a rheological behavior that depends on the viscosities ratio, the surface tension, surfactants, flow-type parameter and the coupled effects of the fluid structure and the kinematics properties of the flow, mainly in the non linear regimen, which is the reason because the dynamics of the fluid particles is an area of current research. As an approach to understand the physical properties of these fluids, studies on the deformation, break-up and coalescence of drops have been performed since the pioneering work of Taylor (1932).

The deformation and breakup of liquid drops is dependent on the kinematics properties of the imposed flow, in particular of the second invariant of the rate of deformation tensor II_{2D} and the flow-type parameter, α (see Astarita, 1979). In experimental studies of the dynamics, break-up and coalescence of drops, then it is necessary to be able to modify on demand the external flow field parameters causing the drop deformation. For example, Taylor (1934) use two flow devices, a Parallel Band Apparatus (PBA) and a Four-Roll Mill (FRM), which manually manipulated each of the rollers speeds to position a drop. Once the drop was in place, Taylor was able to track changes that occur on the drop shape as a function of the imposed flow, although for a short time, until the drop was ejected from its unstable position.

Four-Roll Mills and Parallel Band cover a wide interval in the flow-type parameter; however, there is a gap between them that the flow fields generated by co-rotating Two-Roll Mill (TRM) geometries fill. PBA works for simple shear flow, corresponding to a flow-type parameter $\alpha = 0$; FRM setup is more effective in the interval $0.4 \le \alpha \le 1$ [Yang *et al.* (2001)], whereas TRM is effective in the interval $0.03 \le \alpha \le 0.3$ [Reyes and Geffroy (2000b)].

In the case of the Four-Roll Mill, Bentley and Leal (1986a) have shown how to control -for long times- the position of drops within the flow field generated by a FRM by adjusting with a computer and in real-time the speed of rotation of each cylinder, and maintaining simultaneously constant values for the elongational flow field parameters. This is achieved by modifying the angular velocity of the cylinders, which displaces the stagnation point to a new position, in order to drive the motion of the drop along the stable flow direction towards the stagnation point. Leal's research group at the Chemical Engineering Department at UC Santa Barbara had worked since that with the FRM apparatus, mainly in the pure elongational flow regimen. In the case of the Parallel Band Apparatus, Birkhoffer (2005) proposes computer-controlled flow cell based on the PBA using a digital proportional-integral-differential controller.

We present a nonlinear control applied to study the rheology of a drop in an elongational flow field with vorticity. Large deformations on fluid particles such as drops occur typically in regions containing saddle points. However, the kinematics of a saddle-point flow does not allow for long observation times of the deformed drop due to the outgoing streamlines, which advect the drop away from the flow region of interest. Thus, it is necessary to accurately control the centroid of the drop to study its desired rheology. A suitable control mechanism is essential to maintain the drop position under known flow conditions for times that are long compared to the intrinsic time scales of the dynamics.

In this work, the control mechanism of the position of a drop around the stagnation point of the flow generated by a co-rotating Two-Roll Mill is presented. This control is based on the Poincaré-Bendixson theorem for two-dimensional ordinary differential equations. Namely, when a particle moves within a closed region containing a saddle point inside and the vector field of the equation points inwards at the boundary of the region, the particle undergoes a stable attractive periodic motion. We show that given a prescribed tolerance region, around the unstable stagnation point, an incoming flow can always be generated when the center of mass of the drop reaches the boundary of the tolerance region. This perturbed flow is produced by adjusting the angular velocity of the cylinders, calculated using an analytical solution for the flow, without significant change in the flow parameters. This gives the time dependent analogue of the Poincaré-Bendixson situation just described. The drop is controlled in a perturbed attracting periodic trajectory around the saddle point, while being confined to a prescribed tolerance area. This mechanism is very different from the one used for the proportional control which modifies the unstable nature of the saddle point by adjusting the angular velocities of the cylinder to project the motion along the stable direction only. In the proposed control the effect of the unstable direction combined with the flow readjustment produces the periodic motion.

A nonlinear control strategy, based on the Poincaré-Bendixson theory, is proposed and studied. In essence, the control generates the planar motion of a particle (the centroid in this case) to ensure a periodic motion of the drop centroid inside a prescribed area around the saddle point. In addition, a numerical study and an experimental situation are presented to illustrate the effect of the control on the drop motion. The implementation of the nonlinear control is within a closed region containing the saddle point, with the velocity field pointing inward at every point on the boundary, while the particle undergoes a stable attractive periodic motion. Thus, given a prescribed tolerance region around the stagnation point, it is always possible to generate a controlled incoming flow whenever the center of mass of the liquid drop reaches the boundary of the tolerance region to force the centroid back into the prescribed tolerance region.

The proposed control scheme is capable of keeping the values of the global flow parameters within a small tolerance, and redirecting the liquid-drop centroid toward the stagnation point on a time scale that is much shorter than that of the evolution, with minimal impact upon the liquid-drop dynamics.

We also performed detailed studies of the sensitivity of the control to imperfections in the shape of the geometry that generates the flow, and the variation of the velocity in the servomotors which control the position of the stagnation point. Also a detailed comparison between numeric and experimental data is also presented. This provides a complete study of the two dimensional situation. Finally, we present some open questions both in the modeling and in the theory. In particular in the possibility of extending the current results to drops of complex fluids such as viscoelastic drops, vesicles, capsules, and other immersed objects.

Given that there are no detailed studies available on the influence of the control scheme upon the drop's forms, we study numerically the influence of this control on the motion of a two dimensional drop by solving the Stokes equations in a container subjected to the appropriate boundary conditions on the cylinders and the free surface of the drop. These equations are solved with the Boundary Element Method for a variety of flows and drop parameters in order to study the perturbation effects introduced by the application of the control scheme and to provide the appropriate parameters for future experimental studies. In particular, an easy to implement control scheme is an essential tool prior to undertakings any experimental studies of the drop's dynamics -under elongational flows with vorticity at small Capillary numbers-, for large deformation of drops capsules and other objects, as well as for break up and coalescence of embedded objects.

The proposed method is simple. As the drop evolves under flow conditions, its centroid is tracked. When the drop drifts away and its centroid overtakes the prescribed domain about the nominal stagnation point, the flow is modified by adjusting the angular velocity of the cylinders according to the values obtained from the approximate solution for flows generated by TRMs. Essentially, by adjustment the angular velocities of the cylinders, the outgoing streamline environment is change into an incoming one, reversing the direction of motion of the drop, which is now towards the nominal stagnation point along a stable direction. The reversal of direction does not alter significantly the deformation rates applied upon the drop; thus, the drop's dynamics is essentially undisturbed. The process is repeated as needed and the drop is confined for long times under steady and known conditions.

This study is complementary to the previous one (Bentley 1986a, Bentley 1986b) because it allows detailed studies of the time evolution of the drop's parameters -mainly its deformation and the orientation. In contrast to FRMs, the numerical results presented allow us to study the dynamics of embedded objects under the nominal flow conditions when the drop remains at the stagnation point, as well as the drop in the "controlled flow" with the corresponding parameters. The study was carried out both under several viscosity ratios and various geometric setups, assessing the robustness of the proposed method and the very small influence it has on the drop behavior. The influence of the control scheme on the drop's parameters is small with respect to the nominal flow (around 1%). As well, the proposed control scheme is capable of relocating the stagnation point on a time scale much shorter than the time scale of the drop's evolution. Moreover, this control would remain effective during times much longer than the internal time scales for the drop evolution.

The fact that a two-dimensional numerical model for the drop is used does not compromise the results here presented, mainly because the control scheme should be applicable for small deformation of drops or for drops with very small trajectories about the stagnation point. For highly elongated drops, capsules and other complex objects, the simulated values of the drop's shape may differ from the experimental values, but the applicability of the control scheme should be equally robust.

2. Formulation of the control problem

As already demonstrated by Bentley (1986a), the only way to maintain fixed the position of the drop with respect to the flow field is by changing the location of the stagnation point via adjustments of the angular velocities of the cylinders, with the constraint that these changes must avoid significant modifications of the flow field. Consequently, a useful control scheme for flows by TRMs or FRMs has to maintain the drop as close as possible to the stagnation point for a sufficiently long time, making possible studies of the drop dynamics. From now on, the selected flow field conditions of a TRM are called nominal, and its properties such as the shear rate, flow-type parameter and the position of the stagnation point will be denoted by the subscript Nom.

Figure 1 shows the streamlines around the stagnation point of the unperturbed flow field generated by a co-rotating TRM. When a liquid or rigid particle is placed around the unstable stagnation point, the particle drifts in the direction of the outgoing streamlines. The objective of the control is to maintain a drop about the stagnation point of the nominal flow conditions. To construct the control scheme, the trajectory of the centroid of the drop is analyzed as the solution of a two dimensional dynamical system. In this case we assume that the centroid is away but near the unstable saddle point. Now if the vector field is oncoming on the boundary of a box surrounding the unstable stagnation point the system

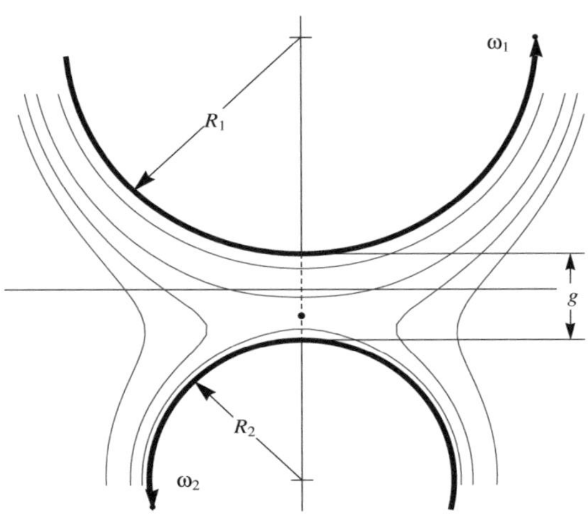

Fig. 1. Streamlines generated by a co-rotating Two-Roll Mill with different radii, showing the stagnation point in the gap between the rollers, g. The position of the stagnation point along the vertical can be moved changing the angular velocities of the rollers. The value of the flow-type parameter is a function of the position of the stagnation point.

will settle into a periodic orbit inside the box provided the vector field is time independent. When there exists time dependence, a bounded motion (which can be periodic or quasi-periodic) is produced inside the box. This motion is equally robust as the time independent case, confining the centroid of the drop to any prescribed region provided the appropriate incoming flow can be produced at the boundary.

In Fig. 2a, a rectangle is shown about the stagnation point of the flow field. The boundaries of this rectangle serve as the limits where the position of the center of mass of the fluid particle is allowed to stay at the nominal flow conditions. Fig. 2b shows a detailed sketch of the tolerance domain *PQRS* above and next to the nominal stagnation point. The dark streamlines correspond to the nominal flow field with outgoing flow exiting through the side QO_R and entering at the side RO_L of the tolerance area. The dashed flow lines which correspond to the shifted stagnation point y_{SS} with the flow field essentially reversed: the flow enters the *PQ* side and exists via *RS*. Also the flow lines which correspond to $-y_{SS}$ behave in a symmetric manner relative to the symmetric tolerance area.

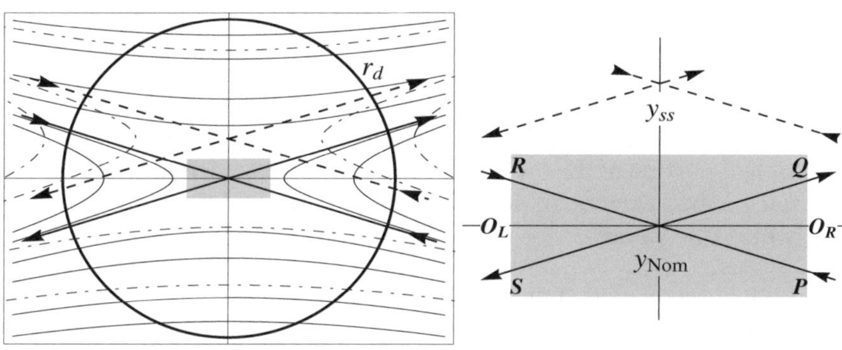

(a) (b)

Fig. 2. Streamlines around the stagnation point of the unperturbed flow field generated by a TRM.

When a fluid particle is placed around the stagnation point and the flow is started, the particle's centroid drifts away. Assume the centroid is initially at the position *A* -at time $t = 0$- located inside the tolerance area shown in Fig. 3. In this position, the flow corresponds to the nominal conditions, and the centroid is subsequently advected along the outgoing direction reaching *B* at $t = t_{\text{on}}$ when the control is applied. The control scheme effect is to displace the stagnation point to y_{SS}, switching the flow field at *B* to one towards the nominal stagnation point. As a result, the centroid follows the flow lines along the path *BC*, arriving at *C* at time $t = t_{\text{off}}$. At t_{off} the flow is reset to the nominal conditions and the stagnation point is moved back to the y_{Nom} position; thus, the centroid follows the path *CD*. At *D* the situation is repeated but now shifting the stagnation point to - y_{SS} until the centroid reaches *E* where the stagnation point is shifted back to the nominal value and the centroid moves towards *F* where the process is repeated. It is to be noted that the transit time spent along paths *AB*, *CD*, *EF* is much longer that the transit time along sections *BC* and *DE* because of the small eigenvalues associated with the corresponding incoming directions which produce the motion shown. The slope of l_{in} can be modified in order to adjust the instant when the flow is reset to the nominal conditions.

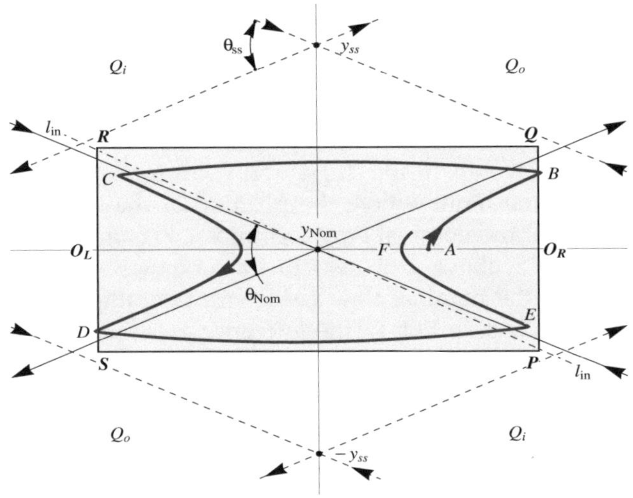

Fig. 3. Trajectory ABCDEF followed by the centroid of a fluid particle in the controlled flow field. The control area is shown in grey. The nominal flow corresponds to the darker continuous lines, and the dashed lines show the relative displacement of the flow field during the controlled portion of the cycle. The angle between the incoming and outgoing streamlines at the nominal stagnation point θ_{Nom} and the angle at the corrective flow θ_{ss} have essentially the same values.

The purpose of the implemented control scheme is to produce always an incoming flow for the drop at the boundary of the tolerance area. Thus the center of mass is effectively moved as a dynamical system with an unstable rest point (the stagnation point) but with an incoming vector field in an area surrounding the box. This arrangement guarantees the existence of an incoming vector field that provides a fixed periodic solution by the Poincaré-Bendixson theory; see Ross (1984). In this case since the incoming vector field is time dependent, a bounded trajectory is obtained that is approximately periodic. This nonlinear procedure of balancing the repulsion at the critical point with the correction of the boundary of the tolerance region always produces a very robust bounded trajectory inside any prescribed area.

All displacements of the stagnation point are assumed to be carried out on a time scale small compared to the dynamics of the drop. In the theoretical description above, both the centroid of the drop and the streamlines are adjusted instantaneously. For a laboratory experiment this will not be the case: the exact position of the centroid is determined after processing the flow field images, determining the centroid position and then the driving motors modify accordingly the flow field within a finite response time. The latter time lags are taken into account in order to calculate the effect of the control in experimental situations.

The relevant times involved are τ_1 associated to the velocity of the video system -fps- and the time of capture and processing of all images, the finite response time τ_2 of the cylinders to readjust their velocity as a consequence of the control, mainly due to the inertia of the mechanical system, and the time of response τ_3 of the fluid around the drop to the adjustment in the velocity of the TRM. The time τ_3 of adjustment of the flow can be estimated as the diffusion time $\tau_3 = l^2/\nu$ based on the gap g between the cylinders and the kinematic viscosity ν. The total response time $\tau_1 + \tau_2 + \tau_3 = \tau_c$ must be smaller than the characteristic time τ_d of the internal motion of the drop which is a function of the Capillary number and the viscosity ratio. In the present work, $\tau_1 + \tau_2 \gg \tau_3$ given the material properties of the fluids and geometry of the setup; Stokes solution implies such time scales as well.

To determine the adjusted velocity field -i.e., reposition the stagnation point position y_{SS} (or $-y_{SS}$) needed to ensure an incoming flow along PQ - a new flow line (assumed straight) is calculated, entering at PQ and ending at y_{SS}; see Fig. 3. This requirement gives y_{SS} as a function of the size of the control area. Parametrizing the flow, relating the position of the stagnation point with the angular velocities of the rollers, ω_1 and ω_2, maintaining II_{2D} constant, the required values of $\omega^c{}_1$ and $\omega^c{}_2$ needed to move the position of the stagnation point to y_{SS} can be calculated.

The actual control step is modeled as follows: When the centroid reaches the boundary of the tolerance area at B in Fig. 3, the angular velocities are readjusted to the calculated values $\omega^c{}_1$ and $\omega^c{}_2$ according to the ramp function

$$\omega^c{}_i(t) \;=\; \omega_{Nom,i}\,\frac{\omega^c{}_i \;-\; \omega_{Nom,i}}{2}\left(1 + tanh\left(\frac{t - t_c}{\tau_c}\right)\right) \qquad (1)$$

where t_c is the time when the centroid of the particle reaches the point B, and i takes the values 1 or 2. The following step at point D is calculated in the same manner. It is remarked that during the control steps the flow parameter II_{2D} is kept constant while the change in α is less than 0.5% for all cases.

This control is different from that of Bentley and Leal. In detail, this control does not aim to stabilize the center of mass of the drop at the stagnation point as in the proportional control. The present control takes advantage of the knowledge of the local flow field and balances the unstable motion at the stagnation point with a time dependent incoming flow at the boundary, giving an effective dynamical system with a periodic or quasi-periodic orbit for the centroid. This is achieved by repositioning the velocity field; thus, producing a bounded motion inside the tolerance area as a result of a balance of instability and the modified flow. In contrast, the control by Bentley and Leal modifies the flow to project the trajectory of the centroid along the stable direction at the nominal stagnation point; essentially, a proportional method attempts to trace the centroid of the drop back to the stagnation point.

3. Experimental setup

The Two-Roll Mill experimental setup is shown in Fig. 4. The setup consists mainly of the Two-Roll Mill flow cell with controlled temperature, the driving system, the optical system and the interface system. The flow cell consist of the main body, a set of rollers of different radii with machining tolerances for cylinder's diameters and gap of less than 5 μm. The parallelism and eccentricity of the rollers axes is limited to less than 5 μm , for a top to bottom distance of 10 cm. The driving system consists of two servomotors Kollmorgen AKM-11B with

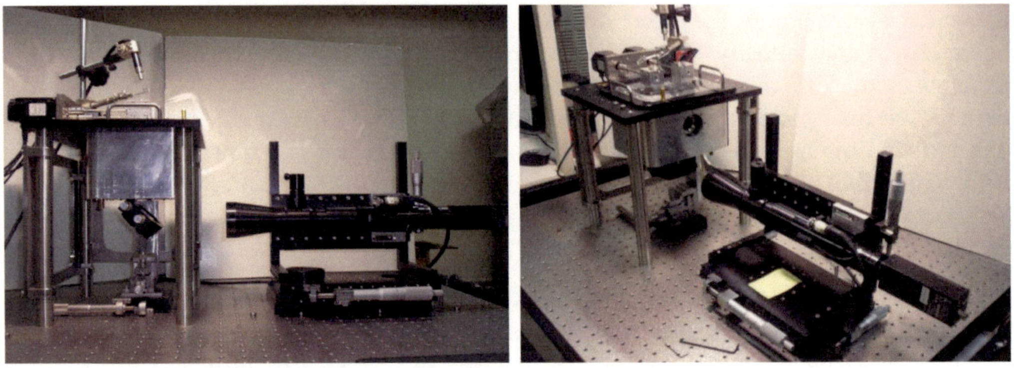

Fig. 4. Experimental setup. The flow cell with the motors and the optical system are shown.

their controllers SERCOS Servostar 300. The optical systems is made up with a Navitar microscope -with a telecentric objective with a motorized magnification of 12X-, an adapter Navitar 1-61390 with a magnification of 2X, and an IEEE 1394 CCD camera, model XCD-X700, by Sony set for a capture rate of 15 and 30 fps. The visual field is about 1.2 mm lengthwise covered with 1024 by 768 pixels. The optical resolution (based on the real part of the Optical Transfer Function) of the full assembly is about 8 μm -or better than 64 line pairs/mm. The main system is mounted on a pneumatically levitated optical workstation by Newport Research.

The typical observed response time for the computer interface, power electronics and cylinder's inertia is less than 0.01 seconds, for changes of rotational speeds less than 5% of the preset values. The flow parameters and the position of the stagnation point are adjusted by varying simultaneously the angular velocities of both cylinders ω_1 and ω_2 keeping II_{2D} constant.

Figure 5 shows a schematic block diagram of the experimental setup. A drop is initially placed near the stagnation point of the flow cell. The optical system captures images of the drop and in the computer, the shape and the center of mass is calculated, and the angular velocities of the rollers are calculated in order to maintain the drop in the position desired. The calculated angular velocities are sent to the motors controllers and a new image of the drop is captured and the cycle is repeated.

The interface system consists of a workstation HP XW4300 with a PCI SERCOS expansion card, which communicates with the motor controllers. The control software is programmed in Visual C++, in real-time mode. A Graphical User Interface (GUI) is used to provide access to the functionality of the application, see Fig. 6. It incorporates two different aspects that can be manipulated separately, one concerning to the video and the other concerning to the control of the motors.

For the video aspect, the GUI window has two displays, one is for the video input, that shows the frame that is acquired by the camera; and the other display is for the processed image, showing the contour of the drop and its center of mass, along with the tolerance area (fixed with the slide bar in the window) and the lines that corresponds to the ingoing and outgoing streamlines of the nominal stagnation point.

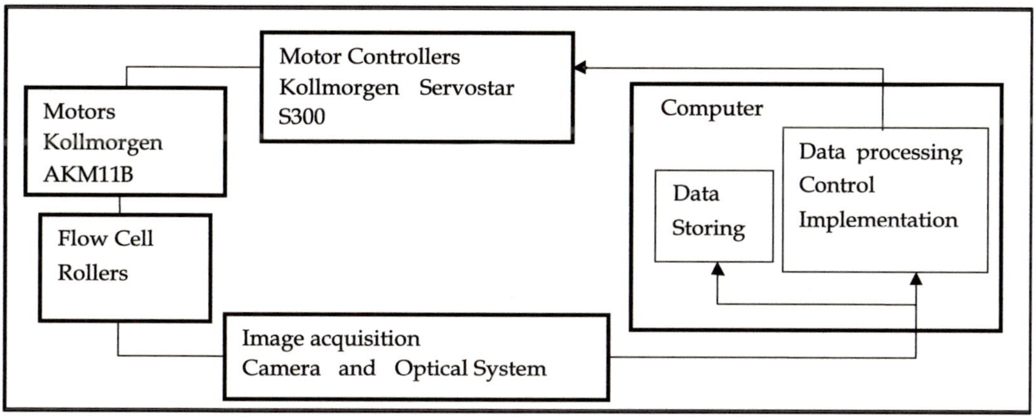

Fig. 5. Experimental block diagram.

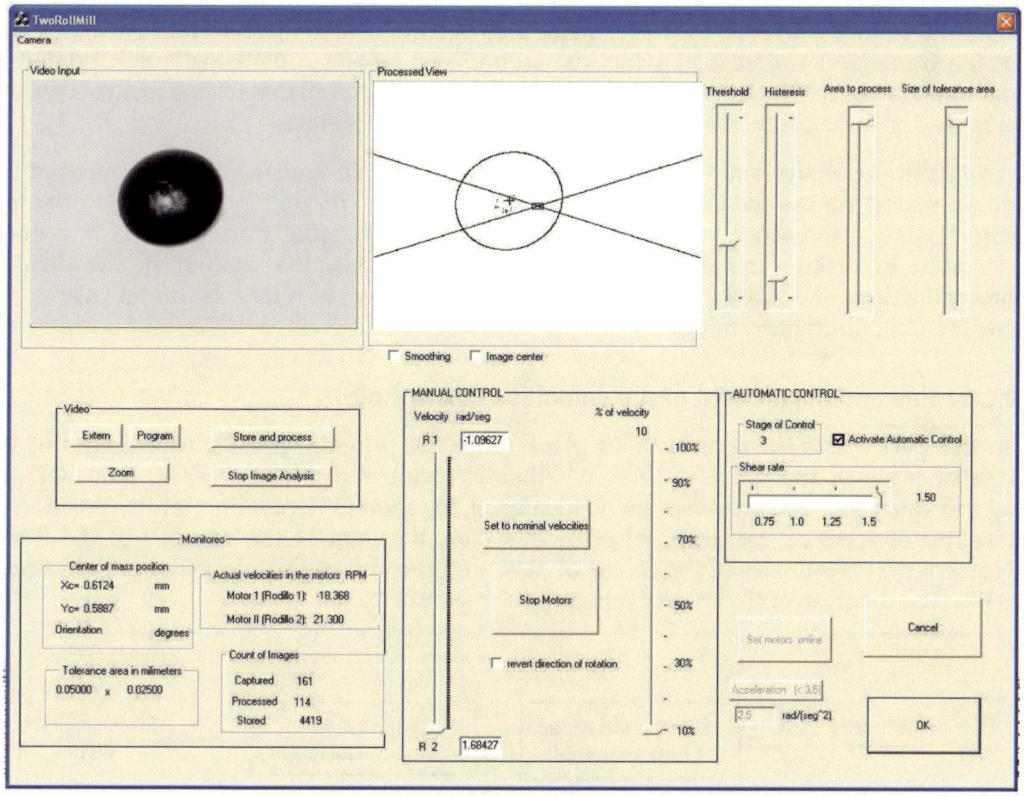

Fig. 6. Graphical User Interface (GUI) of the Two-Roll Mill Experiment.

For the control of the motors, the GUI has two parts, one is for the manual control and the other is for the automatic control. The manual control window is used for the initial positioning of the drop around the nominal stagnation point, inside the tolerance area.

For the control of the motors the window has slide bars that allow controlling the velocities of each cylinder as well as its direction of rotation. In order to place the drops in the center of the image, it is necessary to put it in that location; this is done by manipulating its trajectory with the controls above mentioned. Once the drop is in the right place (it means the center of mass is inside the tolerance area), the automatic control is activated and the experiment goes on. The monitoring section in the window allows us to watch the instantaneous velocities of the motors, the coordinates of the center of mass and the size of the tolerance area.

4. Control scheme implementation

The automatic control for the experiment requires a real-time process. This program consists basically of three parts: (*i*) The image acquisition (*ii*) Image analysis (*iii*) Calculus and adjustment of the velocities of the motors.

4.1 Image acquisition and analysis

The images are provided by the CCD camera using the Instrumental and Industrial Digital Camera Application Programming Interface (*IIDCAPI*) by Sony in the C++ program to handle the frames provided by the camera. This section consists in two parts, the first one take the frame and stored it in a file, the second one creates a list constantly updated in order to always have the last image acquired by the camera available for the analysis section and in case of some delay in acquiring, have a reservoir of frames.

To carry out the analysis of the last frame taken, in order to find the center of mass of the drop (in pixels) we use the Open Source Computer Vision Library (OpenCV). The principal parameter is the threshold value which can be adjusted ranging from 0 to 255 in a gray-scale, used to make a binary image. In the binary image, the contour of the drop is computed using the Canny algorithm. Then the center of mass is found using the corresponding discretized integral.

4.2 Calculus and adjustment of the velocities of the motors

Is in this part where the control takes place. Once the program has the coordinates of the center of mass of the drop, decides whether is inside the tolerance area. If is not, the program calculates and modifies the velocities of the motors depending on the position of the center of mass of the drop. Also, in this part, the data of the number of the frame processed, the coordinates of the center of mass and the size of the tolerance area are stored in a file. The diagram of the control scheme is shown in Fig. 7.

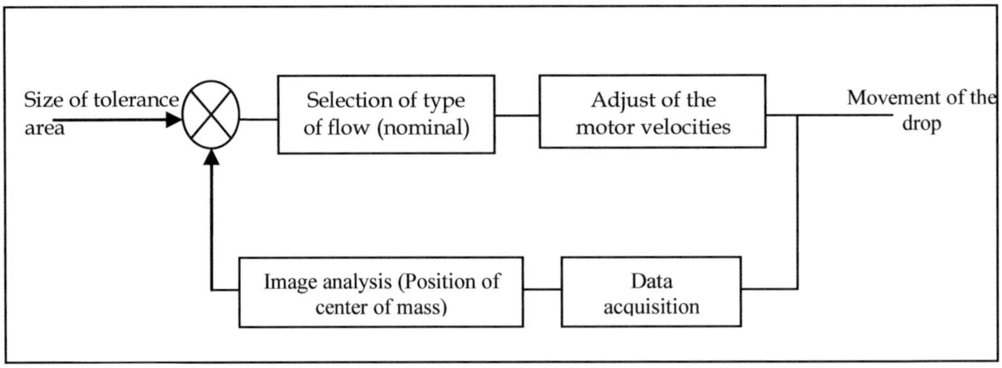

Fig. 7. Diagram of the control scheme.

5. Experimental results

5.1 Parameter used in studies of drop deformation

The parameters that govern the drop deformation and breakup are the ratio of the drop viscosity to that of the suspending fluid λ_μ, the tensorial character of $\nabla\mathbf{u}$, the history of the flow and the initial drop shape.

The Capillary number Ca represents the ratio of flow forces to surface tension, it is given by

$$Ca = \frac{a\,\dot{\gamma}\mu_1}{\Gamma} = \frac{a|\mathrm{II}_{2D}|^{1/2}\mu_1}{\Gamma} \qquad (2)$$

Where II_{2D} is the second invariant of $2\mathbf{D} = \nabla\mathbf{u} + \nabla\mathbf{u}^{\mathrm{T}}$. The tensorial character of $\nabla\mathbf{u}$ gives the flow-type parameter α. For a given elongational flow with vorticity, α values close to 1 imply an elongation dominated flow, while values close to zero imply a flow with strong vorticity; that is α is a measure of the of the strength of the flow causing drops to deform, while the vorticity present in the flow induces s rotation of drops and can inhibit drop breakup. From the definition of the *flow-type parameter* α, (see Astarita, 1979),

$$\frac{1+\alpha}{1-\alpha} = \frac{||\mathbf{D}||}{||\overline{\mathbf{W}}||} \qquad (3)$$

Thus α is given by

$$\alpha = \frac{||\mathbf{D}|| - ||\overline{\mathbf{W}}||}{||\boldsymbol{D}|| + ||\overline{\mathbf{W}}||} \qquad (4)$$

Where $\overline{\boldsymbol{W}}$ is the objective vorticity tensor which measures the rate of rotation of a material point with respect to the rate of deformation's principal axes at that point.

A dimensionless measure of the magnitude of the drop deformation is needed. This parameter is the *Taylor Deformation Parameter* D_T, defined in terms of the longest and shortest semi-axes of the ellipsoidal drop cross section, see Fig 8.

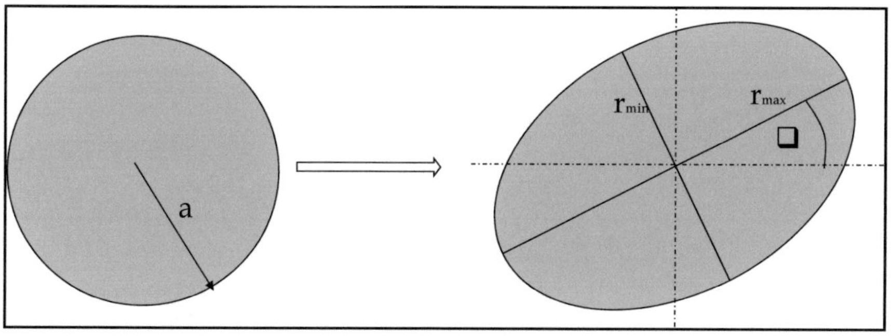

Fig. 8. Deformation Parameter DT and orientation angle.

$$\boldsymbol{D}_T = \frac{r_{max} - r_{min}}{r_{max} + r_{min}} \qquad (5)$$

The orientation angle of the drop is the angle between the longest axis of the drop and the x-axis.

The flow-type parameter value is related with the angles between the incoming and the outgoing axes shown in Fig. 3 as $\theta_{Nom} = 2\,arctan(\sqrt{\alpha_{Nom}})$ and as $\theta_{ss} = 2\,arctan(\sqrt{\alpha_{ss}})$. In the experiment, the outgoing flow direction, about the stagnation point, is always observed responding to small differences of the refractive index of Fluid 1 due to very small differences of temperature between fluid volumes around each cylinder. Based upon reversal of the flow field -counter and co-rotating directions of the cylinders-, the angle between incoming and outgoing flow can be accurately measured within 0.1 degrees inside the visual field -less than 1.2 mm lengthwise. The observed angles are correct within ± 0.05 degrees of the nominal values here presented. This small angular uncertainty implies uncertainties for the nominal flow-type parameter of less than ± 0.5%. Thus, the experimental flow obtained is an excellent approximation for the theoretical one predicted.

The exterior fluid (Fluid 1) is a Polydimethylsiloxane oil DMS 25, η = 485 mPa s, with a relative density of 0.971. A drop (Fluid 2) of vegetable canola oil, filtered through a 3 µm pore size. At 23°C, the viscosity of Fluid 2 is η = 72.6 mPa s, with a relative density of 0.917 is used. Both liquids have a well defined Newtonian behavior at the interval of shear rate values used. The following figures show the effect on the deformation, orientation and trajectories of the centroid of a drop due to variations on the l_{in} parameter. This parameter permit adjusts the control in order to decrease the drifting effects of the τ_c time, which is a nonlinear function of the viscosity of the surrounding fluid. The drop tested had a diameter of 1.0mm, Ca = 0.1031.

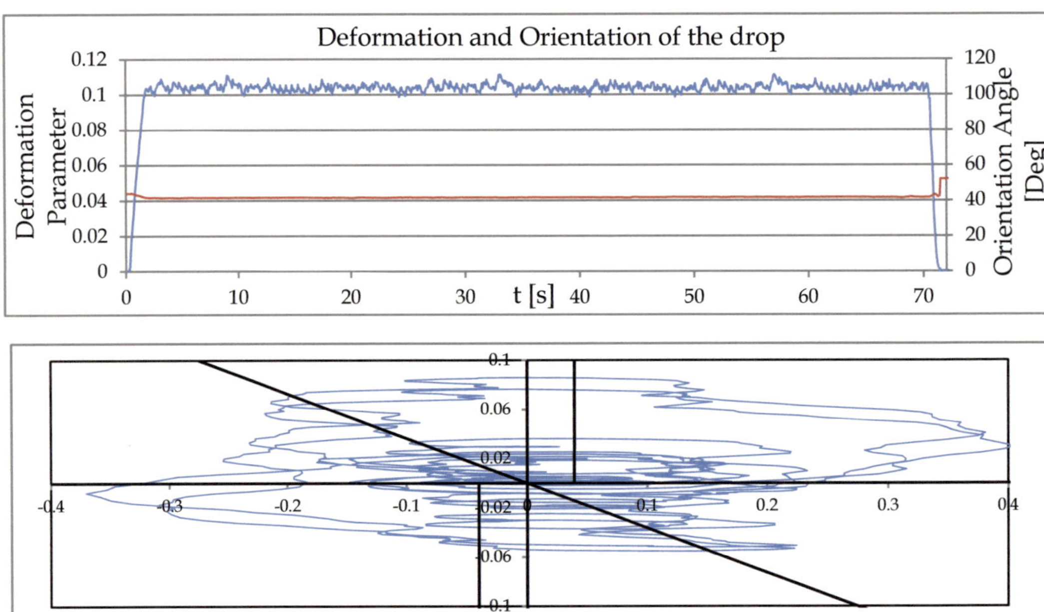

Fig. 9. Deformation, orientation and trajectory of the centroid of the drop, using the parameter l_{in} = 20°. The mean deformation is D_T = 0.1039, STD=0.002 and the mean orientation angle is 41.8°.

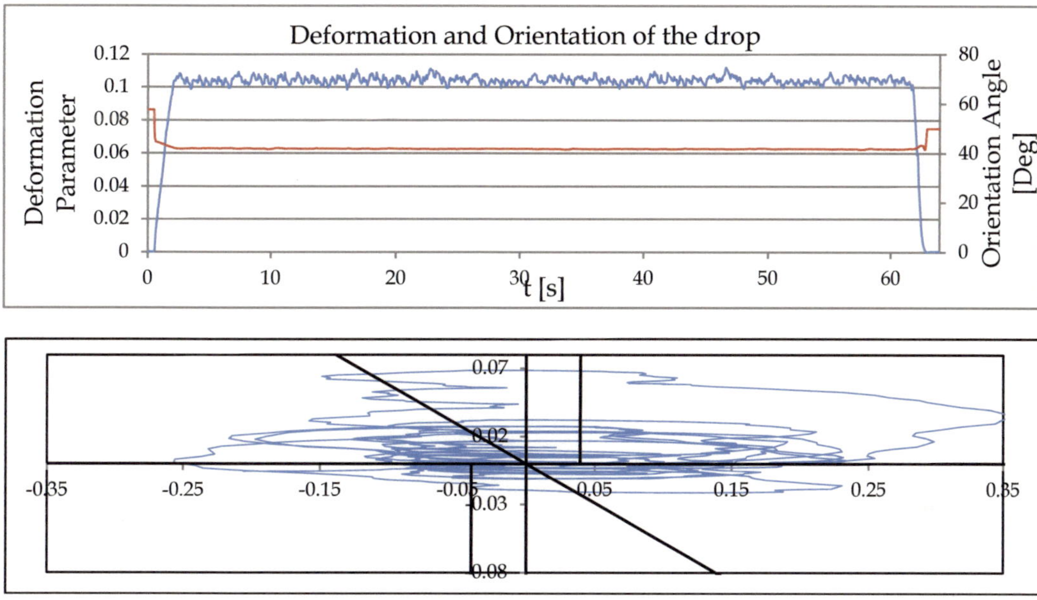

Fig. 10. Deformation, orientation and trajectory of the centroid of the drop, using the parameter l_{in} = 30°. The mean deformation is D_T = 0.1042, STD=0.0021, the mean orientation angle is 41.8°.

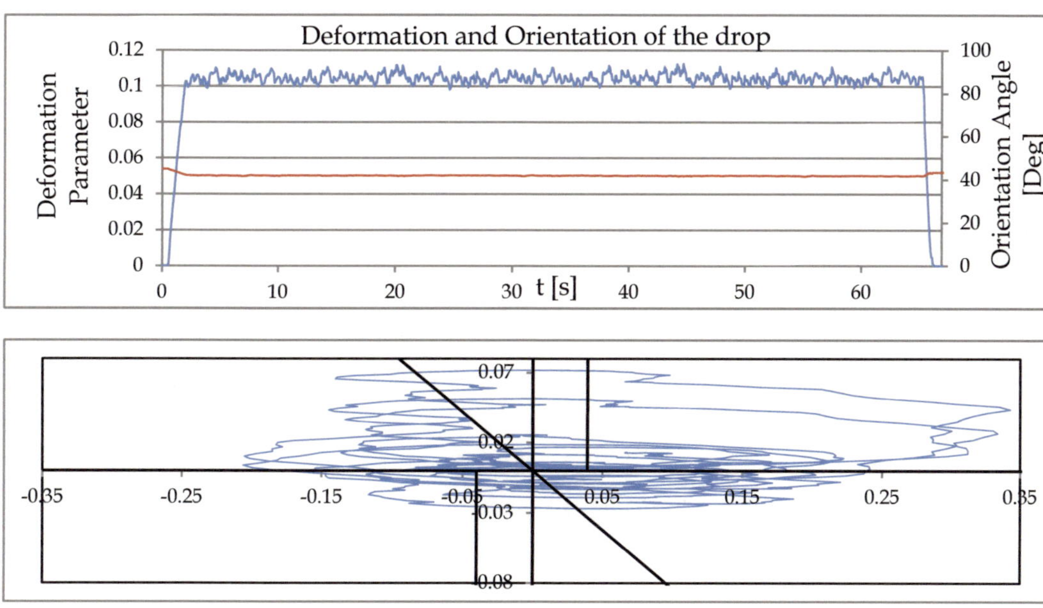

Fig. 11. Deformation, orientation and trajectory of the centroid of the drop, using the parameter l_{in} = 40°. The mean deformation is D_T = 0.1045, STD=0.0025, and the mean orientation angle is 41.8°.

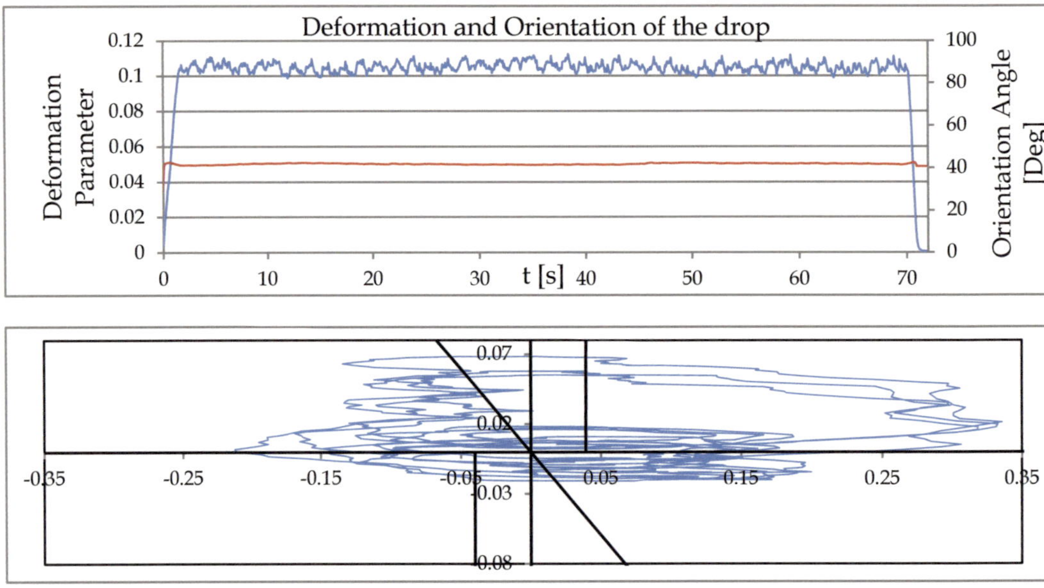

Fig. 12. Deformation, orientation and trajectory of the centroid of the drop, using the parameter $l_{in} = 50^{o}$. The mean deformation is $D_T = 0.1049$, STD=0.0027, the mean orientation angle is 41.8°.

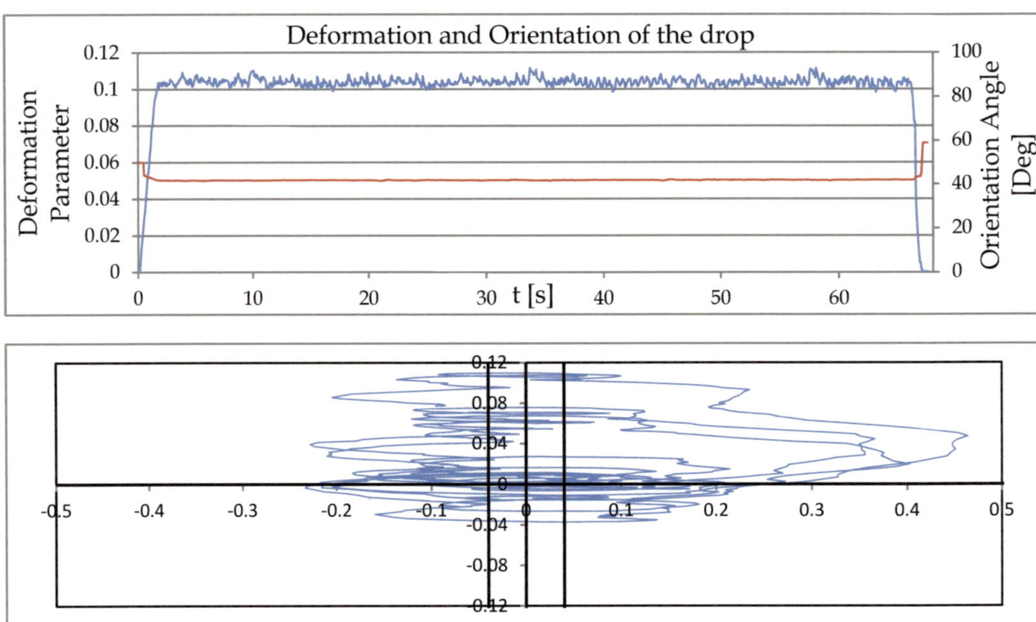

Fig. 13. Deformation, orientation and trajectory of the centroid of the drop, using vertical limits. The mean deformation is $D_T = 0.1039$, STD=0.0020, the mean orientation angle is 41.8°.

6. Numerical results and comparisons with the experimental results

In order to study numerically the evolution of the drop on the proposed control strategy, we use the results of Reyes *et al.* (2011). We just outline the main steps to introduce the extensions used in this Chapter to study the effect of noise and the imperfections present in the experimental situation.

We thus proceed as in Reyes *et al.* (2011) advancing the drop boundary using the equation

$$\frac{\partial \mathbf{x}}{\partial t} = (\mathbf{v}_s \cdot \mathbf{n})\mathbf{n} \qquad (6)$$

where the velocity $\mathbf{v}_s$ in Eq. 6 is determined solving a boundary integral equation on the drop surface (See Pozrikidis, 1992).

The control is implemented as follows: From the new drop surface obtained from Eq. 6, we determined the center of mass of the drop using the same integral of the experiments. Once this is determined, we verified if it falls inside the tolerance region. When it is out of the selected boundary, we applied the correcting flow obtained adjusting the angular velocities using Eq. 1.

We considered the effect of noise and imperfections as follows. In the first place, we noted from the experiments a systematic variation of the angular velocity due to the geometrical imperfections of the cylinder. This was fitted with a single harmonic function with frequency and amplitude determined from the experimental values for the observed flow without drop. The random noise was taken to be white noise. With these new elements, we used the same code described in Reyes *et al.* (2011) to calculate the trajectories of the drop's center of mass, the deformation parameter and the angle of alignment in order to compare with the experimental results of the previous Section.

It is to be noted that the solution of the system are very sensitive in detail to the initial conditions. However, the broad features which depend on the limit cycle nature of the motion for the center of mass are very robust.

Because of this reason, we start the numerical solution with initial conditions for the drop's center of mass which are taken from the experimental data when the initial rapid transients have subsided. Moreover, the numerical flow is started at nominal values since the inhomogeneities presented prevent the analytical construction of the initial flow.

With this, we expect a very good agreement between the numerical and experimental values of the deformation and orientation. This is shown in Figures 14-18. Although those figures were generated with different initial conditions, the broad behavior is similar to the experimental data. The trajectories are also compared. We see good agreement in the broad features, in particular, when the control is operational.

The experimental data shows larger excursions from the nominal stagnation point. These are due to the mismatch between the commercial worm gear and worm mechanism used to reduce the angular velocity of the motors and transmit the motion to the rollers.

In Fig. 19 we display the X and Y component of the motion for the center of mass, the blue lines shows the experimental values and the red lines the numerical solution. The comparison is good considering the mismatch between the initial flows up to t = 10 s.

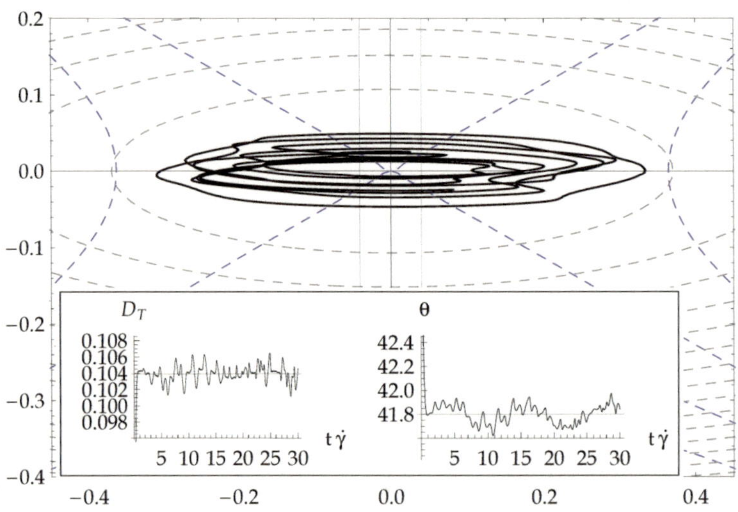

Fig. 14. Trajectory of the centroid of a drop, using vertical limits. The insert graph shows the deformation and the orientation angle. The mean deformation is D_T = 0.1039, STD = 0.000972 with a mean orientation angle of 41.8°, STD = 0.07626.

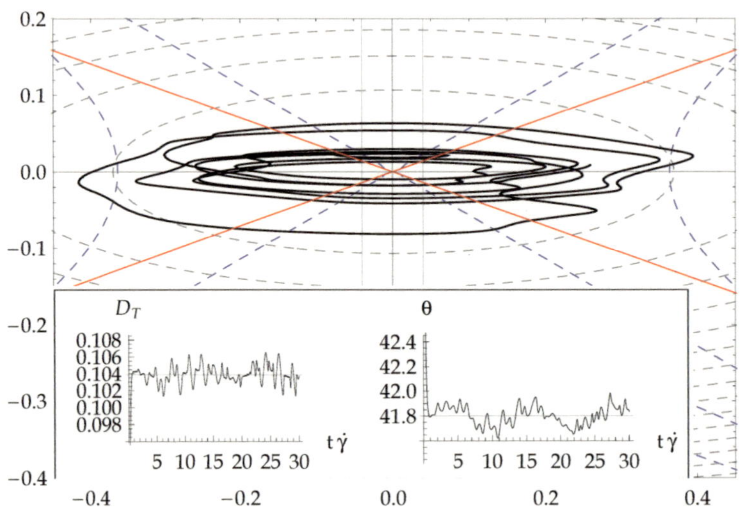

Fig. 15. Trajectory of the centroid of a drop, using the parameter l_{in} same as the incoming axis (19.435 °). The insert graph shows the deformation and the orientation angle. The mean deformation is D_T = 0.10388, STD=0.00104017 with a mean orientation angle of 41.801°, STD = 0.0764.

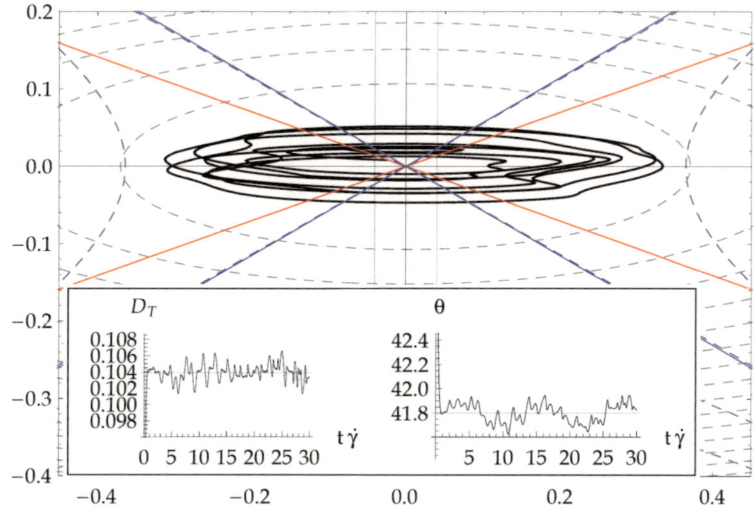

Fig. 16. Trajectory of the centroid of a drop, using the parameter l_{in} = 30 °. The insert graph shows the deformation and the orientation angle. The mean deformation is D_T = 0.103915, STD=0.0009935 with a mean orientation angle of 41.7993°, STD = 0.0784447.

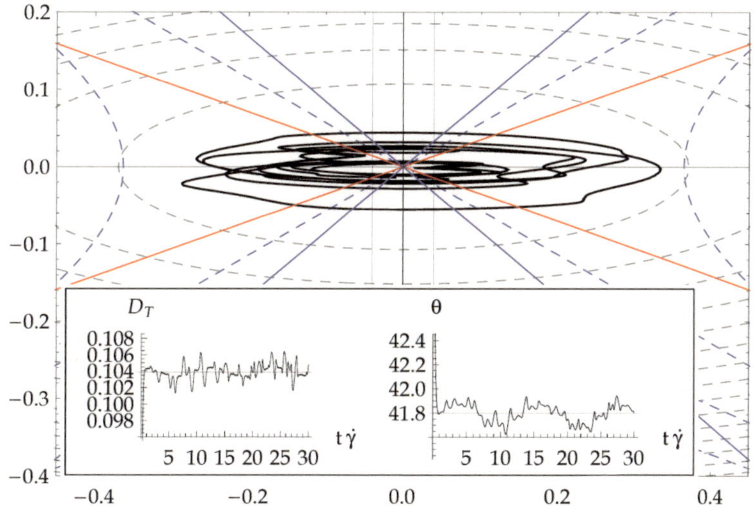

Fig. 17. Trajectory of the centroid of a drop, using the parameter l_{in} = 40 °. The insert graph shows the deformation and the orientation angle. The mean deformation is D_T = 0.103883, STD=0.00092441 with a mean orientation angle of 41.7992°, STD = 0.0735669.

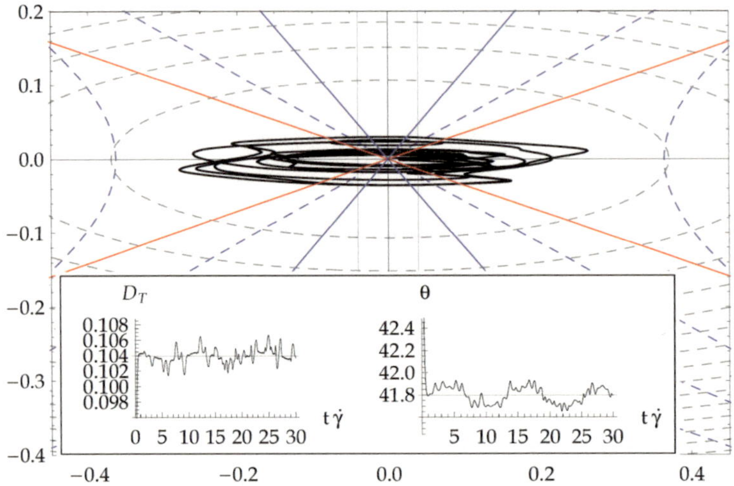

Fig. 18. Trajectory of the centroid of a drop, using the parameter l_{in} = 50°. The insert graph shows the deformation and the orientation angle. The mean deformation is D_T = 0.103934, STD=0.00096193 with a mean orientation angle of 41.7972°, STD = 0.0733545.

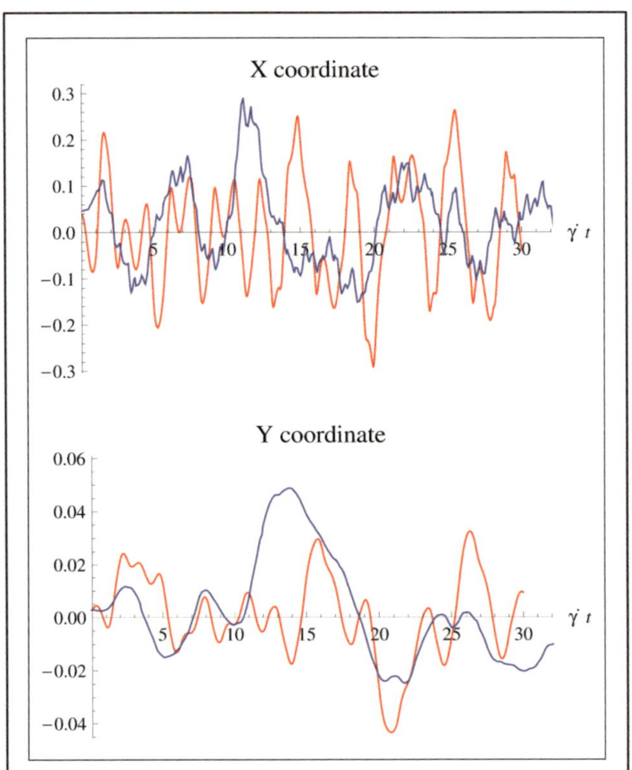

Fig. 19. Experimental and numerical comparisons of the X and Y coordinates of the center of mass of a drop subjected to the control. The blue lines are for the experimental data whereas the red lines are for the numerical data.

Beyond this time, the experimental data shows larger excursions. These are due to the mismatch between the commercial worm gear and worm mechanism used to reduce the angular velocity of the motors and transmit the motion to the rollers.

It is to be noted the remarkable agreement in the actual drop parameters which are the ones of interest.

7. Conclusions

We have shown that it is possible to maintain the position of a drop about the unstable stagnation point of the flow field generated by a TRM setup. This scheme is of the upmost importance for studies of drops in elongational flow with significant vorticity, and of relevance because its space of parameters is not accessible to FRM flows previously studied since Bentley (1986a). Indeed, TRMs expand the family of 2D elongational flows amenable with a four roll mill, albeit the former flows carry amounts of vorticity similar to that of simple shear flows.

But because the TRMs configuration can only displace the stagnation point along the line between the cylinder axes, the control scheme developed for FRMs or PBAs cannot be used for studies of drop's dynamics in elongational flows of the TRM type. For TRMs the control scheme is based upon features of Poincaré-Bendixson limiting cycles. However, these limit cycles do not imply that the wandering trajectory of the drop has to be contained in a very tight region about the nominal stagnation point. The control scheme appears capable of working for a tolerance region with the cases tested experimentally.

The control implemented in the experimental has shown to be successful. At this point, several complications have been seen in the implementation: nevertheless the control scheme is robust enough to keep the drop inside a region where the parameters of interest have a low variation, for a long times, enough to have reliable measures of the relevant parameters. The figures 9-13 shown that despite the trajectory of the drop, the parameters of interest in drop dynamics (D_T and orientation angle) are the same.

It is important to mention that even when the comparisons are made just for only one *flow-type* parameter, the results shown that it is reasonable to expect the same when we will use a different *flow-type* parameter, i.e. a different geometry.

Important differences have been observed between the experimental and numerical trajectories, this disagreement is due to mechanical imperfections, and could be fixed in the future by using an system motion transmission with better precision.

8. Acknowledgment

IYR thanks to CEP-UNAM for funding his graduate fellowship, MAHR thanks DGAPA-UNAM, AAM thanks the FENOMEC program. EG thanks CONACyT research grants.

9. References

Acrivos, A. (1983). The breakup of small drops and bubbles in shear flows. *Ann. NY Acad. Sci.*, 404:1-11

Arfken, G. (1971). *Mathematical Methods for Physicists*. 2nd Ed. Academic Press

Astarita, G. (1979). Objective and generally applicable criteria for flow classification. *Journal of Non-Newtonian Fluids Mechanics*, 6:69-76

Bentley, B. J. and L. G. Leal. (1986a). A computer-controlled four-roll mill for investigations of particle and drop dynamics in two-dimensional linear shear flows. *J. Fluid Mech.* 167:219-240

Bentley, B. J. and L. G. Leal. (1986b). An experimental investigation of drop deformation and breakup in steady two-dimensional linear flows. *J. Fluid Mech.* 167:241-283

Birkhoffer, Beat H. *et al.* (2005). Computer-controlled flow cell for the study of particle and drop dynamics in shear flow fields. *Ind. Eng. Chem. Res.* 44:6999-7009

Geffroy E. and L. G. Leal. (1992). Flow Birefringence of a Concentrated Polystyrene Solution in a Two Roll-Mill 1. Steady Flow and Start-Up of Steady Flow. *J. Polym. Sci. B: Polym. Phys.* 30(12):1329-1349

Gradshteyn, I. S. and I. M. Ryzhik. (1981). *Tables of integrals, series and products*. Academic Press

Jeffery, G. B. (1922). The rotation of two circular cylinders in a viscous fluid. *Proc Royal Society A* 101:169-174

Pozrikidis, C. (1992). *Boundary Integral and singularity methods for linearized viscous flow*. Cambridge University Press

Pozrikidis, C. (2003). *Modeling and simulation of capsules and biological cells*. CRC Press

Rallison J. M. (1980). A note of the time-dependent deformation of a viscous drop which is almost spherical. *J. Fluid Mechanics* 98(3):625-633

Reyes, M. A. H. and E. Geffroy. (2000). Study of low Reynolds number hydrodynamics generated by symmetric co-rotating two-roll mills. *Revista Mexicana de Física* 46(2):135-147

Reyes, M. A. H. and E. Geffroy. (2000). A co-rotating two-roll mill for studies of two-dimensional, elongational flows with vorticity. *Phys. Fluids* 12(10):2372-2376.

Reyes, M. A. H., A. A. Minzoni & E. Geffroy (2011), Numerical study of the effect of nonlinear control on the behavior of a liquid drop in elongational flow with vorticity", *Journal of Engineering Mathematics,* 71(2):185-203.

Ross, S. (1984). *Differential Equations*. Wiley

Singh, P. and L. G. Leal. (1994). Computational studies of the FENE dumbbell model in a co-rotating two-roll mill. *J. Rheol.* 38:485-517

Stone, H. A. (1994). Dynamics of drop deformation and breakup in viscous fluids. *Ann. Rev. Fluid Mech.* 26:65-102

Taylor, G. I. (1932). The Viscosity of a Fluid Containing Small Drops of Another Fluid. *Proc. R. Soc. London*, Ser. A, 138:41-48

Taylor, G. I. (1934). The Formation of Emulsions in Definable Fields of Flow. *Proc. R. Soc. London*, Ser. A 146:501-523

Torza S., R. G. Cox and S. G. Mason. (1972). Particle motions in shear suspensions. *J. Colloid and Interface Sci.* 38(2):395-411

Wang, J. J., D. Yavich, and L. G. Leal. (1994). Time resolved velocity gradient and optical anisotropy in lineal flow by photon correlation spectroscopy. *Phys. Fluids* 6(11):3519-3534

Yang, H., C. C. Park, Y. T. Hu, and L. G. Leal. (2001). The coalescence of two equal--sized drops in a two dimensional linear flow. *Phys. Fluids* 13(5):1087-1106

4

Robust Control Research of Chaos Phenomenon for Diesel-Generator Set on Parallel Connection

Man-lei Huang
School of Automation, Harbin Engineering University, Harbin, China

1. Introduction

Several diesel-generator sets usually operate on parallel connection in ship power system, which has altitudinal nonlinearity. When operating point of system changes, its dynamic property will change markedly. The oscillation phenomenon of ship power system which is acyclic, random and gusty or paroxysmal will occur on light load working condition, it can result in system sectionalizing when it is serious, this phenomenon is called chaos. Chaos is a very complicated phenomenon which is generated by the interaction of each parameter in the nonlinear system. When it appears in ship power system, following the continuous and random oscillation of system operating parameter, which endangers operation security of system seriously, it must be prevented and eliminated effectively in the system. In order to analyze the chaos phenomenon of ship power system, the nonlinear mathematical model of two diesel-generator sets on parallel connection is built in this paper, which reflects the variation law of ship power system. Then the light load working condition of two diesel-generator sets on parallel connection in ship power station is analyzed by using Lyapunov index method on the base of this, seeking the generating mechanism of chaos. A nonlinear robust synthetic controller is designed which is based on the nonlinear mathematical model of diesel-generator set, then a nonlinear robust synthetic control law is developed for the diesel-generator set, it will be applied to control the chaos phenomena, thus providing desirable stability for ship power system.

2. Mathematic model of diesel-generator set on parallel connection

The mathematical model of diesel-generator set include the mathematical model of electromechanical transient process and electromagnetism transient process, first building the mathematical model of electromechanical transient process, then the mathematical

Foundation item: Supported by the National Natural Science Foundation of China under Grant No.60774072; Fundamental Research Funds for the Central Universities of China under Grant No.HEUCFT1005.

model of electromagnetism transient process, the mathematical model of one diesel-generator set is get on the base of this.

The mathematical model of diesel-generator set electromechanical transient process describes the motion law of diesel-generator set[1-3], reflecting the dynamic change process of power angle and angular speed, its expression is

$$\begin{cases} \dfrac{\mathrm{d}\delta}{\mathrm{d}t} = \omega - 1, \\ \dfrac{\mathrm{d}\omega}{\mathrm{d}t} = \dfrac{T_b}{T_a\omega_0}\omega + \dfrac{1}{T_a\omega_0}c_1 + \dfrac{c_2}{T_a\omega_0}L - \dfrac{1}{T_a\omega_0}\dfrac{E'_qU}{X'_d}\sin\delta - \dfrac{1}{T_a\omega_0}\dfrac{U^2}{2}\dfrac{X'_d - X_q}{X'_dX_q}\sin 2\delta. \end{cases} \tag{1}$$

In the equation: δ is power angle of diesel-generator set, ω electric angular speed of diesel-generator set, U terminal voltage of generator stator winding, E'_q q-axis transient electric potential, X reactance of each winding, L output axis displacement of diesel engine govonor actuator, $\omega_0 = 100\pi\mathbf{rad/s}$, T_a, T_b, c_1, c_2 constants, δ, L real values, other variables per unit values.

From Eq.(1) we know, this equation has nonlinear feature.

The mathematical model of diesel-generator set electromagnetism transient process include stator voltage balance equation of generator and electromagnetism transient equation of field winding[3], omitting the effect of damping winding, its expression is

$$\begin{cases} \dfrac{\mathrm{d}E'_q}{\mathrm{d}t} = \dfrac{1}{T_{d0}}E_{fd} - \dfrac{1}{T_{d0}}E'_q - \dfrac{X_d - X'_d}{T_{d0}}I_d, \\ U_d = -RI_d + \omega X_q I_q, \\ U_q = -RI_q - \omega X'_d I_d + \omega E'_q, \\ U = \sqrt{U_d^2 + U_q^2}. \end{cases} \tag{2}$$

In the equation: U is terminal voltage of stator winding, U_d and U_q d-axis and q-axis component of stator winding terminal voltage, R resistance of stator winding, X reactance of each winding, I current of each winding, T time constant of each winding, E'_q q-axis transient electric potential, E_{fd} voltage of exciting winding.

Since E'_q is not easy to measure, we select the terminal voltage of stator winding U as state varible, only needing change E'_q of Eq.(2) into U. According to the relation between varible data, the following form is set up

$$U = E'_q - c_3\delta \tag{3}$$

Substituting Eq.(3) into the first item Eq.(2), we have

$$\frac{\mathrm{d}U}{\mathrm{d}t} = \frac{1}{T_{d0}}E_{fd} - \frac{1}{T_{d0}}U - \frac{c_3}{T_{d0}}\delta - c_3\omega - \frac{X_d - X'_d}{T_{d0}}I_d + c_3 \tag{4}$$

Substituting Eq.(3) into Eq.(1) and combining with Eq.(4), we have

$$\begin{cases} \dfrac{\mathrm{d}\delta}{\mathrm{d}t} = \omega - 1, \\ \dfrac{\mathrm{d}\omega}{\mathrm{d}t} = \dfrac{T_b}{T_a\omega_0}\omega + \dfrac{1}{T_a\omega_0}c_1 + \dfrac{c_2}{T_a\omega_0}L - \dfrac{1}{T_a\omega_0}\dfrac{(U + c_3\delta)U}{X_d'}\sin\delta - \dfrac{1}{T_a\omega_0}\dfrac{U^2}{2}\dfrac{X_d' - X_q}{X_d'X_q}\sin 2\delta, \\ \dfrac{\mathrm{d}U}{\mathrm{d}t} = -\dfrac{c_3}{T_{d0}}\delta - c_3\omega - \dfrac{1}{T_{d0}}U + \dfrac{1}{T_{d0}}E_{fd} - \dfrac{X_d - X_d'}{T_{d0}}I_d + c_3 \end{cases} \tag{5}$$

We know from the expression of current I_d

$$I_d = \frac{E_q' - U\cos\delta}{X_d'} \tag{6}$$

Substituting Eq.(3) into Eq.(6), we have

$$I_d = \frac{U + c_3\delta - U\cos\delta}{X_d'} \tag{7}$$

Substituting Eq.(7) into Eq.(5), we get

$$\begin{cases} \dfrac{\mathrm{d}\delta}{\mathrm{d}t} = \omega - 1, \\ \dfrac{\mathrm{d}\omega}{\mathrm{d}t} = \dfrac{T_b}{T_a\omega_0}\omega + \dfrac{1}{T_a\omega_0}c_1 + \dfrac{c_2}{T_a\omega_0}L - \dfrac{1}{T_a\omega_0}\dfrac{(U + c_3\delta)U}{X_d'}\sin\delta - \dfrac{1}{T_a\omega_0}\dfrac{U^2}{2}\dfrac{X_d' - X_q}{X_d'X_q}\sin 2\delta, \\ \dfrac{\mathrm{d}U}{\mathrm{d}t} = -\dfrac{X_d c_3}{T_{d0}X_d'}\delta - c_3\omega - \dfrac{X_d}{T_{d0}X_d'}U + \dfrac{1}{T_{d0}}E_{fd} + \dfrac{X_d - X_d'}{T_{d0}X_d'}U\cos\delta + c_3 \end{cases} \tag{8}$$

Eq.(8) is the nonlinear mathematical model of one diesel-generator set, which reflects the relationship of interaction and mutual influence among power angle, speed and voltage, describing the variation law of three variables more exactly.

We select *d-q* axis of first synchronous generator as reference frame, building the mathematical model of two diesel-generator sets on parallel connection. Suppose two diesel-generator sets have the same power, type and parameters, *d,q* component of load current is I_d, I_q, power anger difference of two synchronous generators is δ_{12}, the mathematic model of diesel-generator set on parallel connection is

$$\begin{cases} \dfrac{\mathrm{d}\delta_i}{\mathrm{d}t} = \omega_i - 1, \\ \dfrac{\mathrm{d}\omega_i}{\mathrm{d}t} = \dfrac{T_b}{T_a\omega_0}\omega_i + \dfrac{1}{T_a\omega_0}c_1 + \dfrac{c_2}{T_a\omega_0}L_i - \dfrac{1}{T_a\omega_0}\dfrac{(U_i + c_3\delta_i)U_i}{X_d'}\sin\delta_i - \dfrac{1}{T_a\omega_0}\dfrac{U_i^2}{2}\dfrac{X_d' - X_q}{X_d'X_q}\sin 2\delta_i, \\ \dfrac{\mathrm{d}U_i}{\mathrm{d}t} = -\dfrac{X_d c_3}{T_{d0}X_d'}\delta_i - c_3\omega_i - \dfrac{X_d}{T_{d0}X_d'}U_i + \dfrac{1}{T_{d0}}E_{fdi} + \dfrac{X_d - X_d'}{T_{d0}X_d'}U_i\cos\delta_i + c_3 \end{cases} \tag{9}$$

In the equation: $i = 1,\ 2$, subscript 1 indicates the first diesel-generator set, subscript 2 indicates the second diesel-generator set.

Current coupling relation of two diesel-generator sets is

$$\begin{bmatrix} I_d \\ I_q \end{bmatrix} = \begin{bmatrix} I_{d1} \\ I_{q1} \end{bmatrix} + \begin{bmatrix} \cos\delta_{12} & \sin\delta_{12} \\ -\sin\delta_{12} & \cos\delta_{12} \end{bmatrix} \begin{bmatrix} I_{d2} \\ I_{q2} \end{bmatrix} \tag{10}$$

Eq.(10) describes the current distribution relation after two diesel-generator sets enter into parallel connection.

Voltage coupling relation of two diesel-generator sets is

$$\begin{bmatrix} U_{d1} \\ U_{q1} \end{bmatrix} = \begin{bmatrix} \cos\delta_{12} & \sin\delta_{12} \\ -\sin\delta_{12} & \cos\delta_{12} \end{bmatrix} \begin{bmatrix} U_{d2} \\ U_{q2} \end{bmatrix} \tag{11}$$

Eq.(11) describes the voltage restriction relation after two diesel-generator sets enter into parallel connection.

Eq.(9), Eq.(10) and Eq.(11) are the nonlinear mathematical model of two diesel-generator sets on parallel connection, which reflects the relationship of interaction and mutual influence between two diesel-generator sets, describing the variation law of power angle, speed and voltage on two diesel-generator sets exactly.

3. Chaos oscillation analysis of diesel-generator set on parallel connection

This paper will make research on the stability of ship power system by the variation law of power angle, speed and voltage on diesel-generator set. Due to the power transmission between the diesel-generator sets on parallel connection, it is apt to produce power oscillation. Power oscillation is the dynamic process of power of diesel-generator set regulating repeatedly under the effect of some periodic interference.

Power oscillation is a chaos oscillation in nature from the point of view of chaos. The fundamental feature of chaotic motion is highly sensitive to initial condition, the track which is produced by two initial values which is very near each other, will separate according to index pattern as time elapses, Lyapunov index is the quantity which describes the phenomenon. Distinguishing methods of time series chaotic character include fix quantity analysis and ocular analysis, first making numerical analysis for Lyapunov index and judging the condition of chaos emerging, then determining if the chaos exists or not under this condition by the method of ocular analysis. The methods of ocular analysis include time course method, phase path chart method, strobe sampling method, Poincare cross section method and power spectrum method. The methods of calculating Lyapunov index include definition method, Wolf method and Jacobian method, Jacobian method is a method of calculating Lyapunov index which develops in real application. This paper will use Jacobian method to calculate Lyapunov index.

Considering following differential equation system

$$\dot{x} = F(x) \tag{12}$$

In the equation: $\dot{x} = \frac{dx}{dt}$, $x \in R^m$. The evolution of tangent vector e of dot $x(t)$ in the tangent space can be expressed by the equation as follow

$$\dot{e} = T(x(t))e \,, T = \frac{\partial F}{\partial x} \tag{13}$$

In the equation: T is Jacobian matrix of F. The solution of equation can be expressed as

$$e(t) = U(t, e(0)) \tag{14}$$

In the equation $U : e(0) \mapsto e(t)$ is mapping of linear operator. The asymptotic behavior of mapping U can be described by index as

$$\lambda(x(0), e(0)) = \lim_{t \to \infty} \frac{1}{t} \ln \frac{\|e(t)\|}{\|e(0)\|} \tag{15}$$

So, the Lyapunov index of system (12) can be formulated as the mean of above repeat process

$$\begin{aligned} \lambda &= \lim_{k \to \infty} \frac{1}{k\Delta t} \sum_{j=1}^{k} \ln \frac{\|e((j+1)\Delta t)\|}{\|e(j\Delta t)\|} \\ &= \lim_{k \to \infty} \frac{1}{k\Delta t} \ln \frac{\|e((k+1)\Delta t)\|}{\|e(k\Delta t)\|} \frac{\|e(k\Delta t)\|}{\|e((k-1)\Delta t)\|} \cdots \frac{\|e(2\Delta t)\|}{\|e(\Delta t)\|} \end{aligned} \tag{16}$$

For a phase space of n dimension, there will be n Lyapunov index, arranging them according to the order from big to small, supposing $(\lambda_1 \geq \lambda_2 \geq \cdots \geq \lambda_n)$, λ_1 is called maximum Lyapunov index. Generally speaking, having negative Lyapunov index corresponds with contracting direction, the tracks which are near are stable in the part, corresponding periodic motion. The positive Lyapunov index indicates that the tracks which are near separate by index, the strange attractor is formed in phase space, the Lyapunov index λ is bigger, the chaotic nature of system is stronger, vice versa. For a phase space of n dimension, the maximum Lyapunov index is whether bigger than 0 or not is the basis of judging the system if has chaos oscillation or not.

Computing Lyapunov index of two diesel-generator sets on parallel running with light load separately, two diesel-generator sets all use conventional controllers. Fig.1 and Fig.2 give the phase diagram of power angle, speed and voltage of two diesel-generator sets on parallel running with 12.5% load separately.

Two diesel-generator sets run for 100 seconds on parallel connection with 12.5% load, the initial value of No.1 diesel-generator set is: $(\delta, \omega, U) = (0.1017, 1.0662, 0.9762)$, Lyapunov index is: $\lambda_1 = 0.076789$, $\lambda_2 = 0.035235$, $\lambda_3 = -0.197558$; the initial value of No.2 diesel-generator set is: $(\delta, \omega, U) = (0.1022, 1.0662, 0.9762)$, Lyapunov index is: $\lambda_1 = 0.076806$, $\lambda_2 = 0.035230$, $\lambda_3 = -0.197571$.

Fig.3 and Fig.4 give the phase diagram of power angle, speed and voltage of two diesel-generator sets on parallel running with 25% load separately.

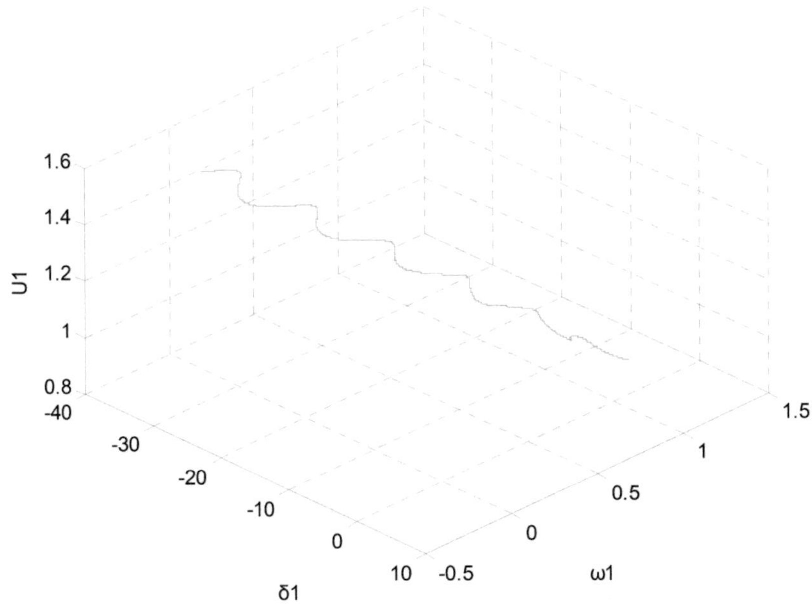

Fig. 1. Phase diagram of No.1 diesel-generator set when two sets load 12.5% on parallel connection

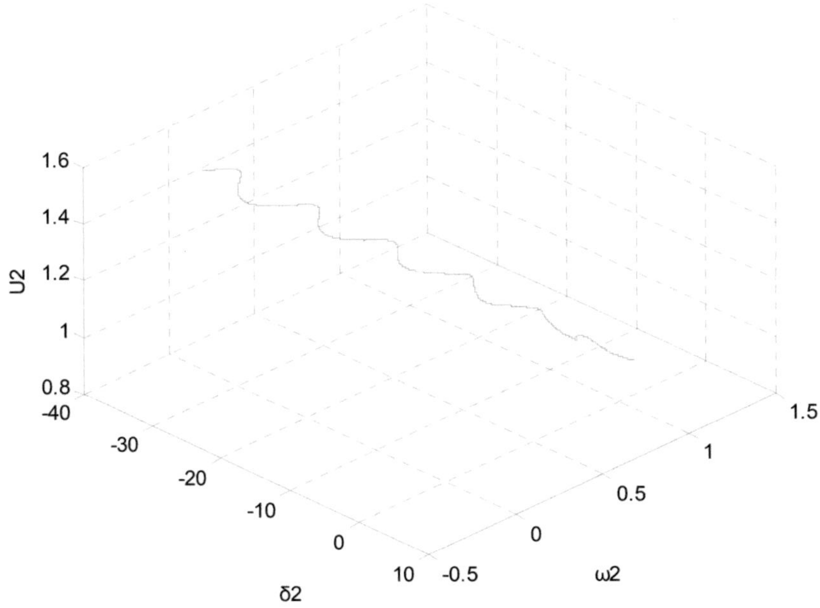

Fig. 2. Phase diagram of No.2 diesel-generator set when two sets load 12.5% on parallel connection

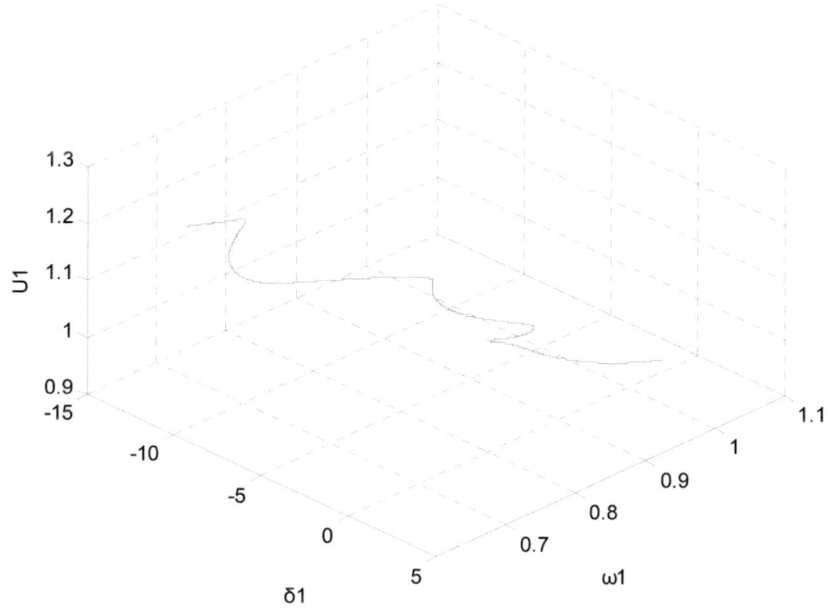

Fig. 3. Phase diagram of No.1 diesel-generator set when two sets load 25% on parallel connection

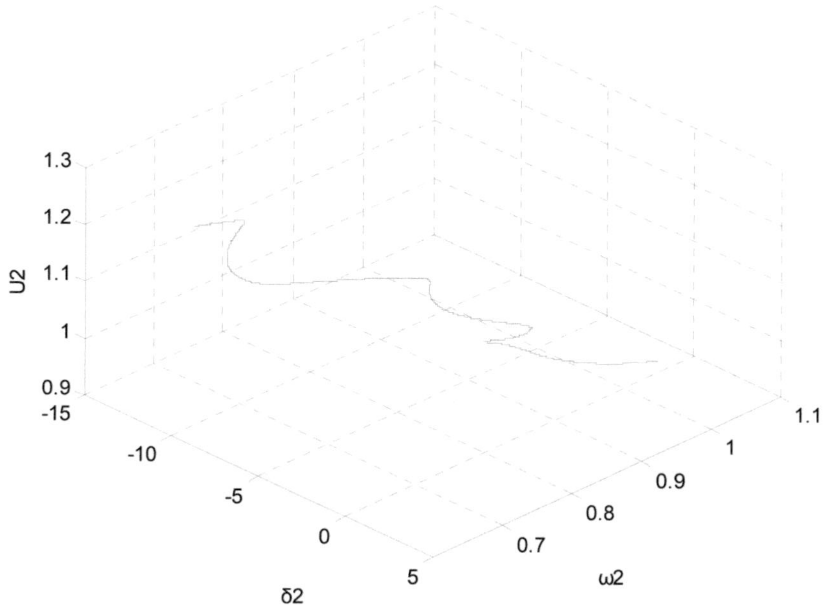

Fig. 4. Phase diagram of No.2 diesel-generator set when two sets load 25% on parallel connection

Two diesel-generator sets run for 100 seconds on parallel connection with 25% load, the initial value of No.1 diesel-generator set is: $(\delta,\omega,U)=(0.1812,1.0607,0.9686)$, Lyapunov index is: $\lambda_1=0.079251$, $\lambda_2=0.034251$, $\lambda_3=-0.199955$;the initial value of No.2 diesel-generator set is: $(\delta,\omega,U)=(0.1820,1.0607,0.9686)$, Lyapunov index is: $\lambda_1=0.078311$, $\lambda_2=0.034953$, $\lambda_3=-0.199742$.

From Fig.1 to Fig.4 we can see, all maximum Lyapunov indexes of the system are greater than 0, showing that the system exists chaotic phenomenon. Two diesel-generator sets with light load on parallel connection, enter into chaotic state after running a length of time, their specific expression are the oscillation of power angle and speed. Two diesel-generator sets on parallel connection load the lighter, the oscillation of power angle and speed is severer. The oscillation of power angle means the oscillation of power, the reason is the nonlinearity of ship power system and the power transmission between the two diesel-generator sets. The controller in this paper is proportional controller, which is a linear controller. It can't control the nonlinear character of ship power system, it can't average the load in parallel operation control, there is a power angular difference between the diesel-generator sets, thus engendering the power transmission between the sets, which results in the happening of chaotic phenomenon.

Fig.5 and Fig.6 give the phase diagram of power angle, speed and voltage of two diesel-generator sets on parallel running with 25% load plus periodicity load separately. The periodicity load usually appears in ship power system, it is widespread.

Two diesel-generator sets run for 100 seconds on parallel connection with 25% load increasing periodicity load 0.01sint, the initial value of No.1 diesel-generator set is: $(\delta,\omega,U)=(0.1812,1.0607,0.9686)$, Lyapunov index is: $\lambda_1=0.073257$, $\lambda_2=0.031824$, $\lambda_3=-0.191497$; the initial value of No.2 diesel-generator set is: $(\delta,\omega,U)=(0.1820,1.0607,0.9686)$, Lyapunov index is: $\lambda_1=0.073393$, $\lambda_2=0.030161$, $\lambda_3=-0.189995$.

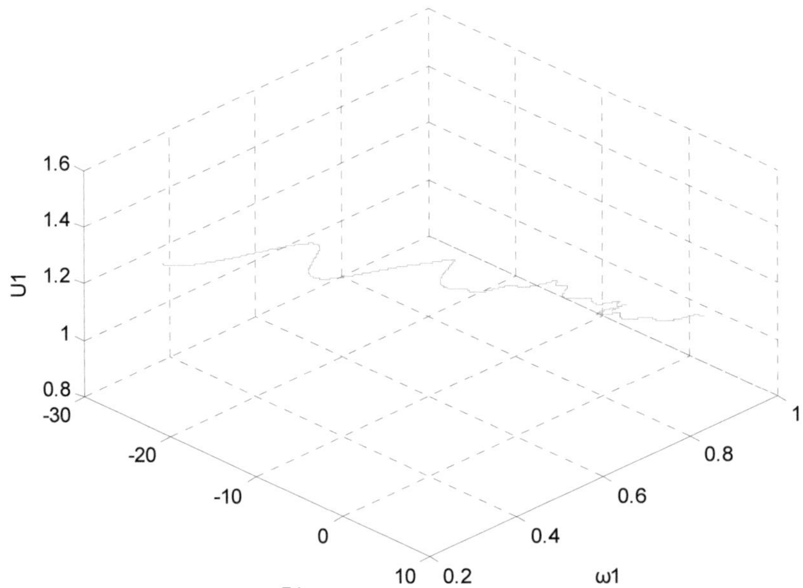

Fig. 5. Phase diagram of No.1 diesel-generator set when two sets increase periodicity load meanwhile load 25% on parallel connection

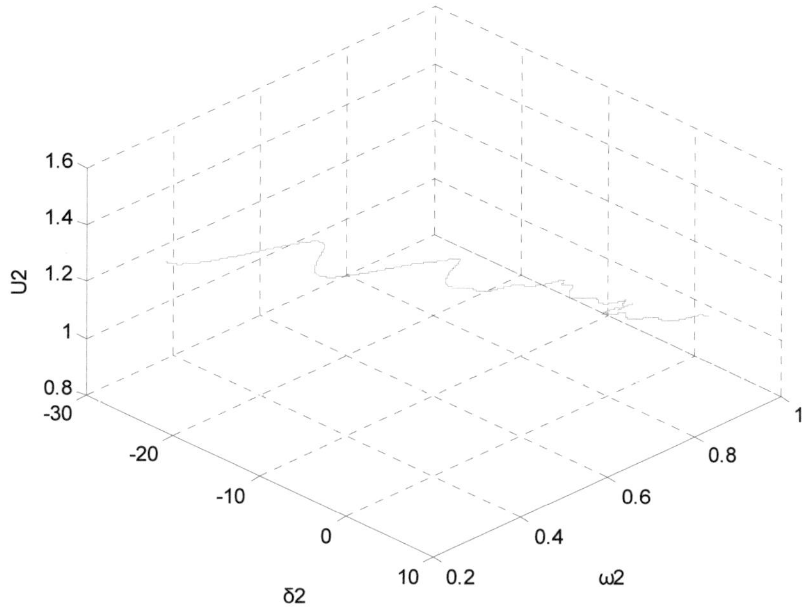

Fig. 6. Phase diagram of No.2 diesel-generator set when two sets increase periodicity load meanwhile load 25% on parallel connection

From Fig.5 and Fig.6 we can see, all maximum Lyapunov indexes of the system are greater than 0, showing that the system exists chaotic phenomenon. The oscillation of power angle and speed in two diesel-generator sets is severer than the state of not increasing periodicity load. Because the periodicity load is nonlinear, increasing periodicity load intensifies the nonlinearity of system, thus aggravating the power oscillation of system[4-6].

The computer simulation results show that it exists chaotic oscillation phenomenon when two diesel-generator sets run on parallel connection with light load. Primary cause of emerging this phenomenon is the nonlinearity of ship power system, minor cause is the power transmission between the two diesel-generator sets. Besides this, using conventional linear controller is also a key factor of generating the chaotic oscillation of system. Only using nonlinear controller, making the nonlinear characteristic of ship power system offset and compensate, can we solve the problem of system chaotic oscillation fundamentally. The chaotic oscillation phenomenon is transition state between stable state and unstable state, it must be prevented in order to ensure the stability of system.

4. Design of nonlinear robust synthetic controller

Mixed H-two/H-infinity control theory is a robust control theory that has speed development from the eighties of 20 century, which can solve the problem of robust stability and robust performance[7-10]. Because diesel-generator set control system is a nonlinear control system, using the method of direct feedback linearization to linearize the nonlinear system, the state feedback controller is designed for linearization system using mixed H-two/H-infinity control theory, thus acquiring nonlinear robust control law in order to reach the purpose of restraining the chaotic phenomenon of ship power system, improving the stability of ship power system.

Because of coupling action between speed and voltage, the nonlinear robust synthetic controller is designed for diesel-generator set in order to control speed and voltage synthetically, making the both interaction in minimum range, thus improving the stability of frequency and voltage in ship power system.

The principle diagram of diesel-generator set synthetic control system based on nonlinear robust synthetic controller is shown in Fig.7. The diesel-generator set synthetic control system is made up of diesel engine, generator, nonlinear robust synthetic controller, actuator, oil feeding mechanism and exciter. Nonlinear robust synthetic controller is made up of two parts, one is nonlinear H-two/H-infinity speed controller, another is nonlinear H-two/H-infinity voltage controller.

The differential equation of actuator is

$$\frac{\mathbf{d}L}{\mathbf{dt}} = -\frac{L}{T_1} + \frac{K_1}{T_1}u_1 \tag{17}$$

The differential equation of exciter is

$$\frac{\mathbf{d}E_{fd}}{\mathbf{dt}} = -\frac{E_{fd}}{T_2} + \frac{K_2}{T_2}u_2 \tag{18}$$

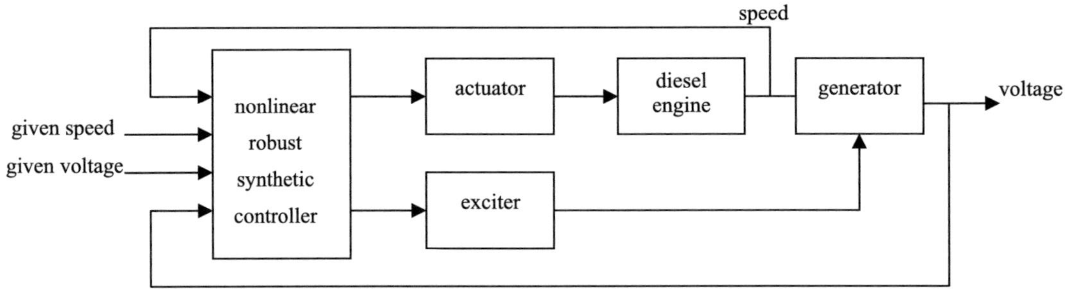

Fig. 7. Principle diagram of diesel-generator set synthetic control system

First step, we design nonlinear H-two/H-infinity speed controller.

Combining Eq.(1) with Eq.(17), we can get the nonliear mathemaical model of diesel engine speed regulation system.

$$\begin{cases} \dfrac{\mathbf{d}\delta}{\mathbf{d}t} = \omega - 1, \\ \dfrac{\mathbf{d}\omega}{\mathbf{d}t} = \dfrac{T_b}{T_a\omega_0}\omega + \dfrac{1}{T_a\omega_0}c_1 + \dfrac{c_2}{T_a\omega_0}L - \dfrac{1}{T_a\omega_0}\dfrac{E'_qU}{X'_d}\sin\delta - \dfrac{1}{T_a\omega_0}\dfrac{U^2}{2}\dfrac{X'_d - X_q}{X'_dX_q}\sin 2\delta, \\ \dfrac{\mathbf{d}L}{\mathbf{dt}} = -\dfrac{L}{T_1} + \dfrac{K_1}{T_1}u_1 \end{cases} \tag{19}$$

Since Eq.(19) has nonlinear feature, the method of direct feedback linearization is used to linearize Eq.(19). Order $x_1 = \delta$, $x_2 = \omega - 1$,

$x_3 = \frac{T_b}{T_a\omega_0}\omega + \frac{1}{T_a\omega_0}c_1 + \frac{c_2}{T_a\omega_0}L - \frac{1}{T_a\omega_0}\frac{E_q'U}{X_d'}\sin\delta - \frac{1}{T_a\omega_0}\frac{U^2}{2}\frac{X_d' - X_q}{X_d'X_q}\sin 2\delta$, so Eq.(19) can be written as

$$\begin{cases} \dot{x}_1 = x_2 \\ \dot{x}_2 = x_3 + d_1 w \\ \dot{x}_3 = \frac{T_b}{T_a\omega_0}x_3 + \frac{c_2 K_1}{T_a T_1 \omega_0}u_1 - \frac{c_2}{T_a T_1\omega_0}L - \frac{E_q'U}{T_a X_d'\omega_0}\cos\delta(\omega - 1) - \frac{U^2(X_d' - X_q)}{T_a X_d' X_q \omega_0}\cos 2\delta(\omega - 1) \end{cases} \tag{20}$$

In the equation: $d_1 w$ is the undesired signal which is assumed for using H-two/H-infinity control method, including equivalence disturbance which is generated by disturbance torque and modeling error.

Assigning virtual controlled variable

$$v = \frac{c_2 K_1}{T_a T_1\omega_0}u_1 - \frac{c_2}{T_a T_1\omega_0}L - \frac{E_q'U}{T_a X_d'\omega_0}\cos\delta(\omega - 1) - \frac{U^2(X_d' - X_q)}{T_a X_d' X_q\omega_0}\cos 2\delta(\omega - 1) \tag{21}$$

So Eq.(20) can be changed as

$$\begin{cases} \dot{x}_1 = x_2 \\ \dot{x}_2 = x_3 + d_1 w \\ \dot{x}_3 = \frac{T_b}{T_a\omega_0}x_3 + v \end{cases} \tag{22}$$

Eq.(22) can be written as matrix form

$$\dot{x} = Ax + B_1 w + B_2 v \tag{23}$$

In the equation: $x = \begin{bmatrix} x_1 \\ x_2 \\ x_3 \end{bmatrix}$, $A = \begin{bmatrix} 0 & 1 & 0 \\ 0 & 0 & 1 \\ 0 & 0 & \frac{T_b}{T_a\omega_0} \end{bmatrix}$, $B_1 = \begin{bmatrix} 0 \\ d_1 \\ 0 \end{bmatrix}$, $B_2 = \begin{bmatrix} 0 \\ 0 \\ 1 \end{bmatrix}$.

Defining the evaluation signal of dynamic performance as

$$\begin{cases} z_\infty = C_1 x + D_{11} w + D_{12} v \\ z_2 = C_2 x + D_{21} w + D_{22} v \end{cases} \tag{24}$$

In the equation: $C_1 = \begin{bmatrix} q_{11} & 0 & 0 \\ 0 & q_{12} & 0 \\ 0 & 0 & q_{13} \end{bmatrix}$, $C_2 = \begin{bmatrix} q_{21} & 0 & 0 \\ 0 & q_{22} & 0 \\ 0 & 0 & q_{23} \end{bmatrix}$, $D_{11} = D_{21} = \begin{bmatrix} 0 \\ 0 \\ 0 \end{bmatrix}$, $D_{12} = \begin{bmatrix} 0 \\ 0 \\ r_1 \end{bmatrix}$,

$D_{22}=\begin{bmatrix}0\\0\\r_2\end{bmatrix}$, C_1, C_2, D_{11}, D_{12}, D_{21}, D_{22} are weighting matrix, $q_{ij}>0$(i=1,2; j=1,2,3)and $r_i>0$ (i=1,2) weighting coefficient. We can select optimal performance combination through changing weighting coefficient, including stability of ship power system, frequency regulation precision and low energy loss of speed regulation system.

For the control system made up of Eq.(23) and Eq.(24), requring design a controller F, making the closed loop system asymptotically stable, moreover H-infinity norm of closed loop transfer function $T_\infty(s)$ from w to z_∞ not more than a given upper bound, in order to ensure the closed loop system have robust stability to uncertainty enter from w; meanwhile making H-2 norm of closed loop transfer function $T_2(s)$ from w to z_2 as small as possible, so as to assure the system performance measured by H-2 norm in a good level, this control problem is called H-two/H-infinity control problem.

From Eq.(23) and Eq.(24), we can get the augmentation controlled object based on mixed H-two/H-infinity control theory

$$P=\left[\begin{array}{c|cc}A & B_1 & B_2\\ \hline C_1 & D_{11} & D_{12}\\ C_2 & D_{21} & D_{22}\end{array}\right] \tag{25}$$

controller F can be solved by corresponding augmentation controlled object P.
For controlled object P, existing H-two/H-infinity state feedback controller:

$$v=Fx=[f_1\ \ f_2\ \ f_3]\cdot\begin{bmatrix}x_1\\x_2\\x_3\end{bmatrix}=f_1x_1+f_2x_2+f_3x_3 \tag{26}$$

In the equation: F is feedback coefficient, which can be got by using μ-Analysis and Synthesis Toolbox in the MATLAB.

Combining Eq.(21) with Eq.(26), we get

$$u_1=\frac{1}{K_1}L+\frac{T_aT_1\omega_0}{c_2K_1}Fx+\frac{E_q'UT_1}{c_2K_1X_d'}\cos\delta(\omega-1)+\frac{U^2T_1(X_d'-X_q)}{c_2K_1X_d'X_q}\cos2\delta(\omega-1) \tag{27}$$

That is nonlinear H-two/H-infinity speed control law of diesel-generator set. Substituting x_1, x_2, x_3 into Eq.(27), we can get practical form of nonlinear H-two/H-infinity speed control law:

$$\begin{aligned}u_1=&\frac{1}{K_1}L+\frac{T_aT_1\omega_0}{c_2K_1}f_1\delta+\frac{T_aT_1\omega_0}{c_2K_1}f_2(\omega-1)+\frac{T_bT_1}{c_2K_1}f_3\omega+\frac{T_1}{c_2K_1}f_3c_1+\frac{T_1}{K_1}f_3L-\frac{T_1}{c_2K_1}\frac{E_q'U}{X_d'}f_3\sin\delta-\\&\frac{T_1}{c_2K_1}\frac{U^2}{2}\frac{X_d'-X_q}{X_d'X_q}f_3\sin2\delta+\frac{E_q'UT_1}{c_2K_1X_d'}\cos\delta(\omega-1)+\frac{U^2T_1(X_d'-X_q)}{c_2K_1X_d'X_q}\cos2\delta(\omega-1)\end{aligned} \tag{28}$$

Second step, we design nonlinear H-two/H-infinity voltage controller.

Combining Eq.(1), Eq.(4) with Eq.(18), we can get the nonliear mathemaical model of synchronous generator voltage regulation system.

$$\begin{cases} \dfrac{\mathrm{d}E_{fd}}{\mathrm{d}t} = -\dfrac{E_{fd}}{T_2} + \dfrac{K_2}{T_2}u_2 \\ \dfrac{\mathrm{d}U}{\mathrm{d}t} = \dfrac{1}{T_{d0}}E_{fd} - \dfrac{1}{T_{d0}}U - \dfrac{c_3}{T_{d0}}\delta - c_3\omega - \dfrac{X_d - X_d'}{T_{d0}}I_d + c_3 \\ \dfrac{\mathrm{d}\delta}{\mathrm{d}t} = \omega - 1 \\ \dfrac{\mathrm{d}\omega}{\mathrm{d}t} = \dfrac{T_b}{T_a\omega_0}\omega + \dfrac{1}{T_a\omega_0}c_1 + \dfrac{c_2}{T_a\omega_0}L - \dfrac{1}{T_a\omega_0}\dfrac{E_q'U}{X_d'}\sin\delta - \dfrac{1}{T_a\omega_0}\dfrac{U^2}{2}\dfrac{X_d' - X_q}{X_d'X_q}\sin 2\delta \end{cases} \tag{29}$$

We select the error of voltage ΔU as state varible, The relation between U and ΔU is

$$\Delta U = U - U_0 \tag{30}$$

In the equation: U_0 is initial value of stator winding terminal voltage, its value is 1.

We use P_e repalce $\dfrac{E_q'U}{X_d'}\sin\delta + \dfrac{U^2}{2}\dfrac{X_d' - X_q}{X_d'X_q}\sin 2\delta$, regarding P_e as external disturbance, substituting Eq.(30) into Eq.(29), we get

$$\begin{cases} \dfrac{\mathrm{d}E_{fd}}{\mathrm{d}t} = -\dfrac{E_{fd}}{T_2} + \dfrac{K_2}{T_2}u_2 \\ \dfrac{\mathrm{d}\Delta U}{\mathrm{d}t} = \dfrac{1}{T_{d0}}E_{fd} - \dfrac{1}{T_{d0}}\Delta U - \dfrac{c_3}{T_{d0}}\delta - c_3\omega + (c_3 - \dfrac{1}{T_{d0}}) - \dfrac{X_d - X_d'}{T_{d0}}I_d \\ \dfrac{\mathrm{d}\delta}{\mathrm{d}t} = \omega - 1 \\ \dfrac{\mathrm{d}\omega}{\mathrm{d}t} = \dfrac{T_b}{T_a\omega_0}\omega + \dfrac{1}{T_a\omega_0}c_1 + \dfrac{c_2}{T_a\omega_0}L - \dfrac{1}{T_a\omega_0}P_e \end{cases} \tag{31}$$

Eq.(31) can be written as matrix form

$$\dot{x}' = A'x' + B_1'w' + B_2'u_2 \tag{32}$$

In the equation: $x' = \begin{bmatrix} E_{fd} \\ \Delta U \\ \delta \\ \omega \end{bmatrix}$, $A' = \begin{bmatrix} -\dfrac{1}{T_2} & 0 & 0 & 0 \\ \dfrac{1}{T_{d0}} & -\dfrac{1}{T_{d0}} & -\dfrac{c_3}{T_{d0}} & -c_3 \\ 0 & 0 & 0 & 1 \\ 0 & 0 & 0 & \dfrac{T_b}{T_a\omega_0} \end{bmatrix}$,

$$B_1' = \begin{bmatrix} 0 & 0 & 0 & 0 \\ -\dfrac{X_d - X_d'}{T_{d0}} & 0 & 0 & c_3 - \dfrac{1}{T_{d0}} \\ 0 & 0 & 0 & -1 \\ 0 & \dfrac{c_2}{T_a\omega_0} & -\dfrac{1}{T_a\omega_0} & \dfrac{c_1}{T_a\omega_0} \end{bmatrix},\ B_2' = \begin{bmatrix} \dfrac{K_2}{T_2} \\ 0 \\ 0 \\ 0 \end{bmatrix},\ w' = \begin{bmatrix} I_d \\ L \\ P_e \\ 1 \end{bmatrix}.$$

Defining the evaluation signal of dynamic performance as

$$\begin{cases} z_\infty' = C_1'x' + D_{11}'w' + D_{12}'u_2 \\ z_2' = C_2'x' + D_{21}'w' + D_{22}'u_2 \end{cases} \tag{33}$$

In the equation: $C_1' = \begin{bmatrix} q_{14} & 0 & 0 & 0 \\ 0 & q_{15} & 0 & 0 \\ 0 & 0 & q_{16} & 0 \\ 0 & 0 & 0 & q_{17} \end{bmatrix}, C_2' = \begin{bmatrix} q_{24} & 0 & 0 & 0 \\ 0 & q_{25} & 0 & 0 \\ 0 & 0 & q_{26} & 0 \\ 0 & 0 & 0 & q_{27} \end{bmatrix}, D_{12}' = \begin{bmatrix} r_3 \\ r_4 \\ r_5 \\ r_6 \end{bmatrix}, D_{22}' = \begin{bmatrix} r_7 \\ r_8 \\ r_9 \\ r_{10} \end{bmatrix},$

$D_{11}' = D_{21}' = \begin{bmatrix} 0 & 0 & 0 & 0 \\ 0 & 0 & 0 & 0 \\ 0 & 0 & 0 & 0 \\ 0 & 0 & 0 & 0 \end{bmatrix}$, $C_1', C_2', D_{11}', D_{12}', D_{21}', D_{22}'$ are weighting matrix, q_{ij} >0(i=1,2; j=4-7)and r_i >0(i=3-10)weighting coefficient. We can select optimal performance combination through changing weighting coefficient, including stability of ship power system, voltage regulation precision and low energy loss of excitation system.

From Eq.(32) and Eq.(33), we can get the augmentation controlled object based on mixed H-two/H-infinity control theory

$$P' = \left[\begin{array}{c|cc} A' & B_1' & B_2' \\ \hline C_1' & D_{11}' & D_{12}' \\ C_2' & D_{21}' & D_{22}' \end{array}\right] \tag{34}$$

For controlled object P', existing H-two/H-infinity state feedback controller:

$$u_2 = F'x' = [f_1' \ f_2' \ f_3' \ f_4'] \cdot \begin{bmatrix} E_{fd} \\ \Delta U \\ \delta \\ \omega \end{bmatrix} = f_1'E_{fd} + f_2'\Delta U + f_3'\delta + f_4'\omega \tag{35}$$

That is nonlinear H-two/H-infinity voltage control law of diesel-generator set.

Third step, we design nonlinear robust synthetic controller.

Combining Eq.(28) with Eq.(35), we can get nonlinear robust synthetic controller of diesel-generator set[11-18]

$$\begin{cases} u_1 = \frac{1}{K_1}L + \frac{T_a T_1 \omega_0}{c_2 K_1} f_1 \delta + \frac{T_a T_1 \omega_0}{c_2 K_1} f_2(\omega - 1) + \frac{T_b T_1}{c_2 K_1} f_3 \omega + \frac{T_1}{c_2 K_1} f_3 c_1 + \frac{T_1}{K_1} f_3 L - \frac{T_1}{c_2 K_1}\frac{E'_q U}{X'_d} f_3 \sin\delta - \\ \frac{T_1}{c_2 K_1}\frac{U^2}{2}\frac{X'_d - X_q}{X'_d X_q} f_3 \sin 2\delta + \frac{E'_q U T_1}{c_2 K_1 X'_d}\cos\delta(\omega - 1) + \frac{U^2 T_1 (X'_d - X_q)}{c_2 K_1 X'_d X_q}\cos 2\delta(\omega - 1) \\ u_2 = f'_1 E_{fd} + f'_2 \Delta U + f'_3 \delta + f'_4 \omega \end{cases} \tag{36}$$

Eq.(36) considers the coupling function of speed and voltage, which controls both synthetically. It can increase dynamic precision of speed and voltage, improving the stability of ship power system.

5. Results of computer simulation

The key parameters of diesel-generator set control system in this paper are as follow:

The power of diesel-generator set is 1250kW; the rated speed n =1500r/min; the rotary inertia of set J=71.822kg·m^2; the damping coefficient of set D =5.54;the magnetic pole pair number of generator p=2; the rated torque of diesel engine 11.9kN·m; the maximum troke of output axis 10mm.

The rated voltage of synchronous generator is 390V; the rated current 2310A; the power factor 0.8; the rated frequency 50Hz; the exciting voltage of exciter 83V; the exciting current 7.7A.

Designing nonlinear synthetic controller based on mixed H-two/H-infinity control theory, assuming disturbance signal coefficient of Eq.(23) d_1 =0.1, assuming weighting coefficient of Eq.(24) q_{11} =0.002, q_{12} =0.4, q_{13} =0.5, r_1 =0.01, q_{21} =0.002, q_{22} =0.4, q_{23} =0.5, r_2 =0.01.

Corresponding matrix are

$$A = \begin{bmatrix} 0 & 1 & 0 \\ 0 & 0 & 1 \\ 0 & 0 & -0.0014 \end{bmatrix}, B_1 = \begin{bmatrix} 0 \\ 0.1 \\ 0 \end{bmatrix}, C_1 = C_2 = \begin{bmatrix} 0.002 & 0 & 0 \\ 0 & 0.4 & 0 \\ 0 & 0 & 0.5 \end{bmatrix}, D_{12} = D_{22} = \begin{bmatrix} 0 \\ 0 \\ 0.01 \end{bmatrix}.$$

Using LMI toolbox, we get state feedback controller:

$$F = [-0.3582 \quad -35.1042 \quad -157.7578] \tag{37}$$

Assuming weighting coefficient of Eq.(33) q_{14} =0.1, q_{15}=2800, q_{16} =0.1, q_{17} =0.1, q_{24} =0.1, q_{25} =2800, q_{26} =0.1, q_{27} =0.1, $r_3 = r_4 = r_5 = r_6$ =1, $r_7 = r_8 = r_9 = r_{10}$ =1.

Corresponding matrix are

$$A' = \begin{bmatrix} -0.4545 & 0 & 0 & 0 \\ 0.0011 & -0.0011 & -0.0002 & -0.2043 \\ 0 & 0 & 0 & 1 \\ 0 & 0 & 0 & -0.0014 \end{bmatrix}, B'_1 = \begin{bmatrix} 0 & 0 & 0 & 0 \\ -0.0021 & 0 & 0 & 0.2032 \\ 0 & 0 & 0 & -1 \\ 0 & -0.0004 & -0.0028 & 0.0042 \end{bmatrix},$$

$$C_1' = C_2' = \begin{bmatrix} 0.1 & 0 & 0 & 0 \\ 0 & 2800 & 0 & 0 \\ 0 & 0 & 0.1 & 0 \\ 0 & 0 & 0 & 0.1 \end{bmatrix}, D_{12}' = D_{22}' = \begin{bmatrix} 1 \\ 1 \\ 1 \\ 1 \end{bmatrix}.$$

Using LMI toolbox, we get state feedback controller:

$$F' = [-0.027 \quad -697.798 \quad -0.023 \quad -0.025] \tag{38}$$

Substituting Eq.(37) and Eq.(38) into Eq.(36), we get nonlinear robust synthetic controller of diesel-generator set.

Optimization of weighting function is difficult point of H-two/H-infinity control, needing select repeatedly. After each selection, using LMI toolbox to get state feedback coefficient, substituting simulation model, making charactrsitc test in order to get best combination property index. Fig.8 give the block diagram of the system simulation.

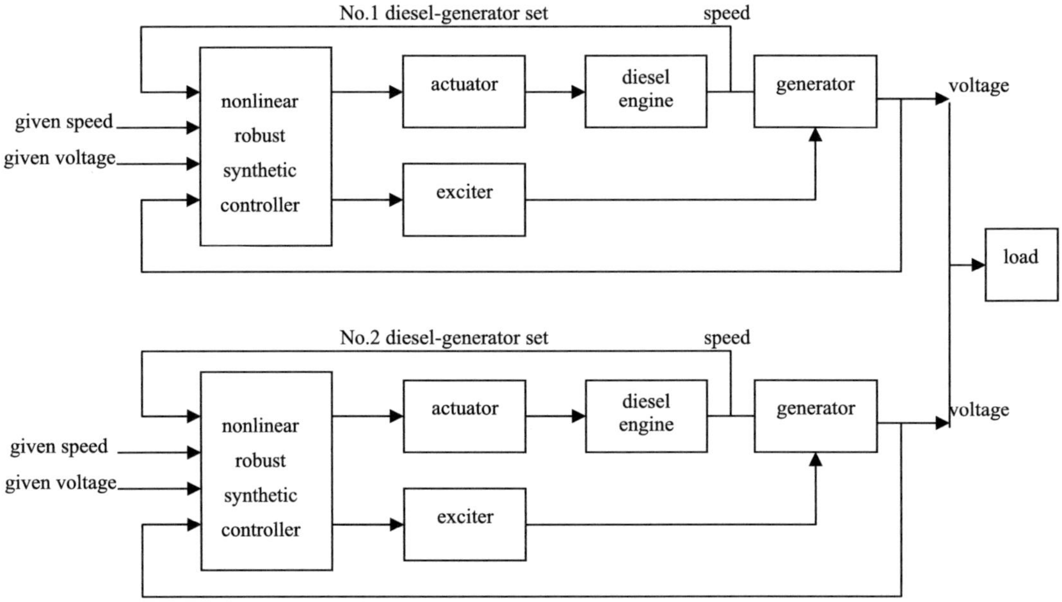

Fig. 8. Block diagram of the system simulation

In order to test and verify the effect of nonlinear robust synthetic controller for restraining the chaotic oscillation of ship power system, Fig.9 and Fig.10 give the phase diagram of power angle, speed and voltage of two diesel-generator sets on parallel running with 12.5% load separately after using nonlinear robust synthetic controller. The initial va1ues of two sets are same as that of generating chaotic oscillation, the running time is also 100 seconds. Calculating the Lyapunov index of two diesel-generator sets on parallel operation, we get the following results. Lyapunov index of No.1 diesel-generator set is: $\lambda_1 = -0.003435$, $\lambda_2 = -0.104389$, $\lambda_3 = -0.230524$; Lyapunov index of No.2 diesel-generator set is: $\lambda_1 = -0.003442$, $\lambda_2 = -0.104382$, $\lambda_3 = -0.230524$.

Fig.11 and Fig.12 give the phase diagram of power angle, speed and voltage of two diesel-generator sets on parallel running with 25% load separately after using nonlinear robust synthetic controller. The initial values of two sets are same as that of generating chaotic oscillation, the running time is also 100 seconds. Calculating the Lyapunov index of two diesel-generator sets on parallel operation, we get the following results. Lyapunov index of No.1 diesel-generator set is: $\lambda_1 = -0.003983$, $\lambda_2 = -0.103822$, $\lambda_3 = -0.230553$; Lyapunov index of No.2 diesel-generator set is: $\lambda_1 = -0.003993$, $\lambda_2 = -0.103811$, $\lambda_3 = -0.230554$.

Fig.13 and Fig.14 give the phase diagram of power angle, speed and voltage of two diesel-generator sets on parallel running with 25% load plus periodicity load separately after using nonlinear robust synthetic controller. The initial values of two sets are same as that of generating chaotic oscillation, the running time is also 100 seconds. Calculating the Lyapunov index of two diesel-generator sets on parallel operation, we get the following results. Lyapunov index of No.1 diesel-generator set is: $\lambda_1 = -0.004331$, $\lambda_2 = -0.103293$, $\lambda_3 = -0.230741$; Lyapunov index of No.2 diesel-generator set is: $\lambda_1 = -0.004341$, $\lambda_2 = -0.103283$, $\lambda_3 = -0.230742$.

From Fig.9 to Fig.12 we can see, all maximum Lyapunov indexes of the system are less than 0, showing that the system doesn't exists chaotic phenomenon and works in a stable range. Two diesel-generator sets with light load on parallel connection, run after using nonlinear robust synthetic controller, their chaotic phenomenon disappears, power angle, speed and voltage run nearby the desired values. It shows that nonlinear robust synthetic controller can control the nonlinear character of ship power system effectively and make the nonlinear characteristic of ship power system offset and compensate, it can solve the problem of system chaotic oscillation fundamentally.

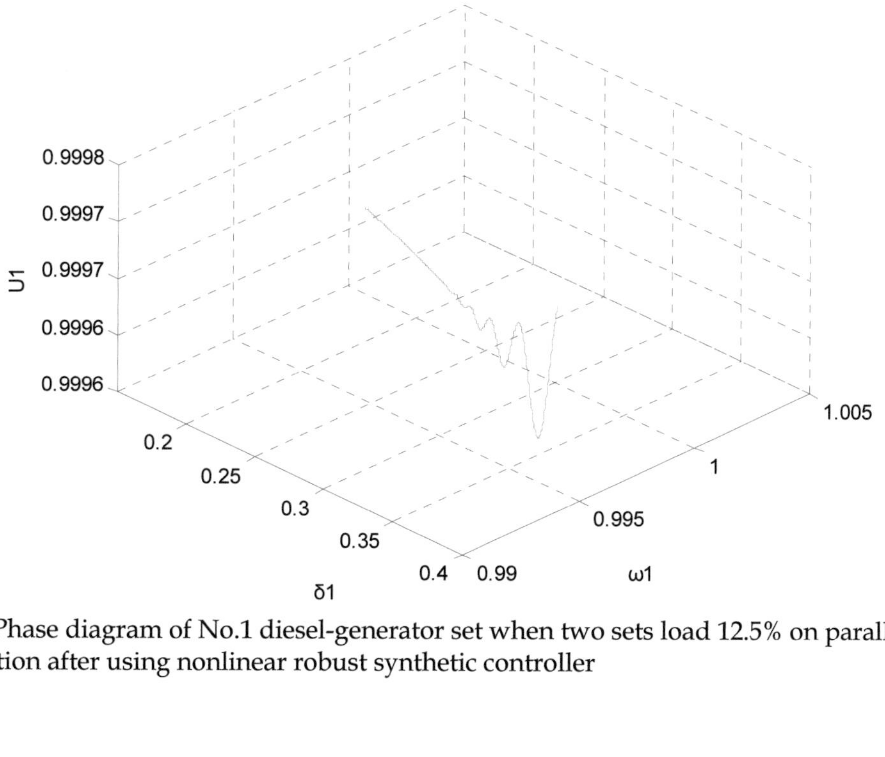

Fig. 9. Phase diagram of No.1 diesel-generator set when two sets load 12.5% on parallel connection after using nonlinear robust synthetic controller

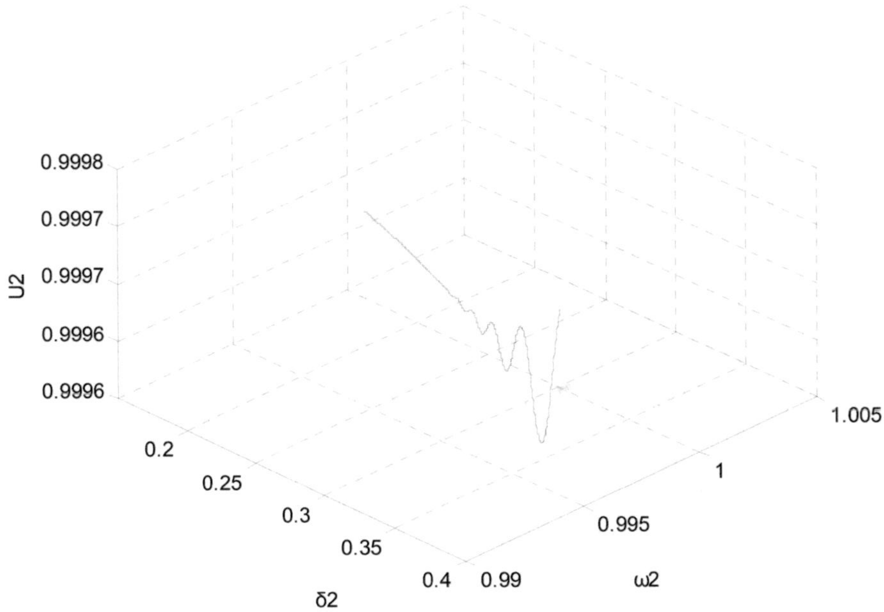

Fig. 10. Phase diagram of No.2 diesel-generator set when two sets load 12.5% on parallel connection after using nonlinear robust synthetic controller

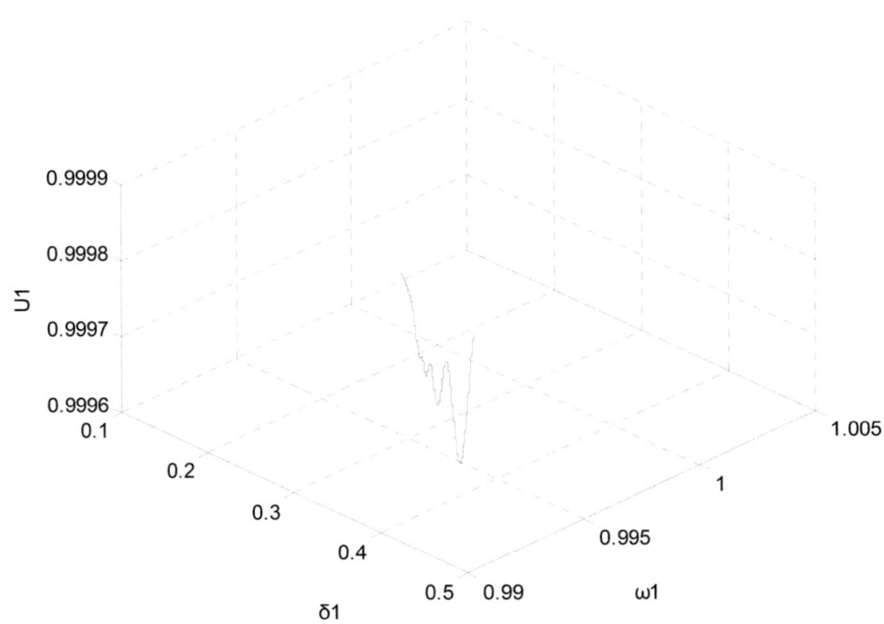

Fig. 11. Phase diagram of No.1 diesel-generator set when two sets load 25% on parallel connection after using nonlinear robust synthetic controller

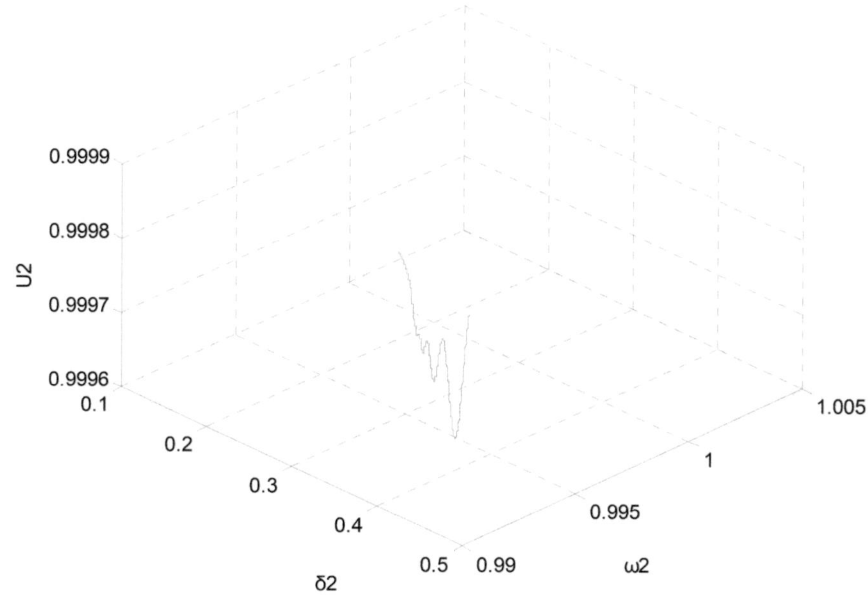

Fig. 12. Phase diagram of No.2 diesel-generator set when two sets load 25% on parallel connection after using nonlinear robust synthetic controller

From Fig.13 to Fig.14 we can see, all maximum Lyapunov indexes of the system are also less than 0, showing that the system also doesn't exists chaotic phenomenon and works in a stable range. Although after adding periodicity load intensifying the nonlinearity of system, nonlinear robust synthetic controller can restrain the chaotic phenomenon of ship power

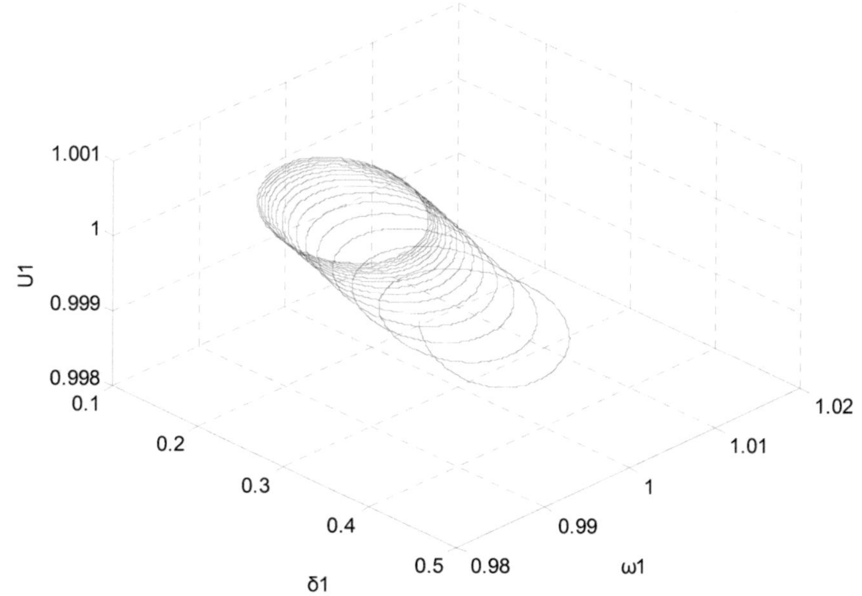

Fig. 13. Phase diagram of No.1 diesel-generator set when two sets load 25% plus periodicity load on parallel connection after using nonlinear robust synthetic controller

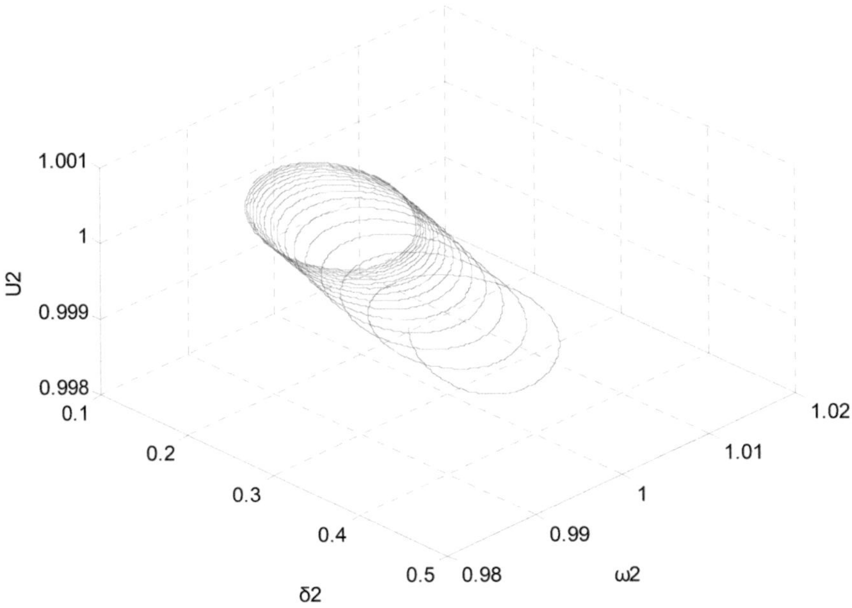

Fig. 14. Phase diagram of No.2 diesel-generator set when two sets load 25% plus periodicity load on parallel connection after using nonlinear robust synthetic controller

system, making the nonlinear characteristic of ship power system offset and compensate, thus improving the stability of ship power system.

Because ship power system made up of diesel-generator sets is a nonlinear system, using nonlinear robust synthetic controller can offset and compensate the nonlinear characteristic of ship power system, it can average the load in parallel operation control, there is no power angular difference between the diesel-generator sets, thus avoiding the power transmission between the sets, solving the problem of system chaotic oscillation fundamentally, improving the voltage and frequency stability of ship power system. The research on nonlinear robust synthetic controller of diesel-generator set provides a new control method for ship power system, having important practical significance and extensive using prospect.

6. Conclusion

In order to analyze the chaos phenomenon of ship power system, the nonlinear mathematical model of two diesel-generator sets on parallel connection is built in this paper, which reflects the relationship of interaction and mutual influence between two sets. The light load working condition of two diesel-generator sets on parallel connection in ship power station is analyzed by using Lyapunov index method, which proves the presence of chaos phenomenon. A nonlinear robust synthetic controller is designed which is based on the nonlinear mathematical model of diesel-generator set. In combining the direct feedback linearization with robust control theory to design the synthetic controller for the diesel-generator set, then a nonlinear robust synthetic control law is developed for the diesel-generator set. The computer simulation results show that the nonlinear robust synthetic controller effectively suppresses the chaos phenomenon of ship power system, thus providing desirable stability for ship power system.

7. References

[1] Huang Manlei, Tang Jiaheng, Guo Zhenming · The Mathematical Model of Diesel Engine Speed Regulation System[J] · Journal of Harbin Engineering University, 1997, 18(6):20~25

[2] Huang Manlei, Li Dianpu, Liu Hongda · Simulation Research on Double-pulse Speed Governor of Diesel Engine[J] · SHIP ENGINEERING, 2002, 24(3):36~38

[3] Huang Manlei, Wang Changhong · Nonlinear mathematical model of diesel-generator set on ship[J] · Journal of Harbin Engineering University, 2006, 27(1):15-19, 47.

[4] Yang Zhengling, Lin Kongyuan · Study on the relation between classical swing equations and chaos[J] · Automation of Electric Power Systems, 2000, 24(7):20~22,45

[5] Jia Hongjie,Yu Yixin,Wang Chengshan · Chaotic phenomena in power systems and its studies [J] · Proceedings of the CSEE, 2001, 21(7):26~30

[6] Wang Baohua, Yang Chengwu, Zhang Qiang. Summary of bifurcation and chaos research in electric power system[J] · Transactions of China Electrotechnical Society, 2005, 20(7):1~10

[7] Bernstein D S, Hadded W M · LQG control with an H_∞ performance bound: A Riccati equation approach · IEEE Trans on Automatic Control, 1989, 34(3):293-305

[8] Zhou K, Glover K, Bodenheimer B, et al · Mixed H_2 and H_∞ performance objectives I: Robust performance analysis · IEEE Trans on Automatic Control, 1994, 39(8):1564-1574

[9] Doyle J, Zhou K, Glover K, et al · Mixed H_2 and H_∞ performance objectives II: Optimal Control · IEEE Trans on Automatic Control, 1994, 39(8):1575-1587

[10] Khargonekar P P, Rotea M A · Mixed H_2/H_∞ control: A convex optimization approach · IEEE Trans on Automatic Control, 1991, 36(7):824-837

[11] Sun Yu-song, Sun Yuan-zhang, Lu Qiang, Shao Yi-xiang · Research on Nonlinear Robust Control Strategy for Hydroelectric Generator's Valve[J] · Proceedings of the CSEE, 2001,21(2):56-59,65

[12] Li Wen-lei, Jing Yuan-wei, Liu Xiao-ping · Nonlinear robust control for turbine main steam valve[J] · Control Theory and Applications, 2003,20(3):387-390

[13] Wang Jin-hua · Design of mixed H_2/H_∞ controller[J] · Control Theory and Applications, 2004,21(1):45-53

[14] Robert Lashlee, Vittal Rao and Frank Kern · Mixed H_2 and H_∞ Optimal Control of Smart Structures · Proceedings of the 33th conference on decision and control, Lake Buena Vista, FL, December 1994:115-120

[15] Curtis P. Mracek and D. Brett Ridgely · Normal Acceleration Command Following of the *F*-16 Using Optimal Control Methodologies: A Comparison

[16] Kap Rai Lee, Do Chang Oh, Kyeong Ho Bang and Hong Bae Park · Mixed H_2/H_∞ Control for Underwater Vehicle with Time Delay and Parameter Uncertainty · Proceedings of the American Control Conference, Albuquerque, New Mexico, June 1997:3225-3229

[17] Kap Rai Lee, Do Chang Oh, Kyeong Ho Bang and Hong Bae Park · Mixed H_2/H_∞ Control with Regional Pole Placement for Underwater Vehicle Systems · Proceedings of the American Control Conference, Chicago, Illinois, June 2000:80-84
[18] Feng Wu, Keat-Choon Goh and Steve Walsh · Robust H_2 Performance Analysis for A High-Purity Distillation Column · Computers Chem. Energy, 1997, 21:8161-8166

5

A Robust State Feedback Adaptive Controller with Improved Transient Tracking Error Bounds for Plants with Unknown Varying Control Gain

A. Rincon[1], F. Angulo[2] and G. Osorio[2]
[1]*Universidad Católica de Manizales*
[2]*Universidad Nacional de Colombia - Sede Manizales - Facultad de Ingeniería y Arquitectura - Departamento de Ingeniería Eléctrica, Electrónica y Computación - Percepción y Control Inteligente - Bloque Q, Campus La Nubia, Manizales, Colombia*

1. Introduction

The design of robust model reference adaptive control (MRAC) schemes for plants in controllable form, comprising unknown varying but bounded coefficients and varying control gain has attracted a great deal of research. Many nonlinear systems may be described by the controllable form; for instance, second order plants (see (Hong & Yao, 2007), (Hsu et al., 2006), (Yao & Tomizuka, 1994), (Jiang & Hill, 1999)) and systems whose nonlinear behavior or part of it, is represented by some function approximation technique (cf. Nakanishi et al. (2005), (Chen et al., 2008), (Tong et al., 2000), (Huang & Kuo, 2001), (Yousef & Wahba, 2009), (Hsu et al., 2006), (Koo, 2001), (Labiod & Guerra, 2007)).

The state adaptive backstepping (SAB) of (Kanellakopoulos et al., 1991) is a common framework for the design of adaptive controllers for plants in controllable form. As is well known, a major difficulty in introducing robustness techniques to SAB based schemes is that the states z_i and the stabilizing functions must be differentiable to certain extent (see (Yao & Tomizuka, 1997), (Yao, 1997), (Ge & Wang, 2003)).

The robust SAB scheme of (Zhou et al., 2004), (Su et al., 2009), (Feng, Hong, Chen & Su, 2008) has the advantage that the knowledge on the upper or lower bounds of the plant coefficients can be relaxed if the controller is properly designed and the control gain is constant or known. The approach is based on the truncation method of (Slotine & Li, 1991), pp. 309. The stabilizing functions are smoothed at each i-th step in order to render it differentiable enough. The following benefits are obtained: i) the scheme is robust with respect to unknown varying but bounded coefficients, ii) upper or lower bounds of the plant coefficients are not required to be known, and iii) the tracking error converges to a residual set whose size is user–defined.

The specific case of unknown varying control gain is an important issue, more difficult to handle than other unknown varying coefficients. The varying control gain is usually handled by means of robustness techniques (cf. (Wang et al., 2004), (Huang & Kuo, 2001), (Bechlioulis

& Rovithakis, 2009), (Li et al., 2004)) or the Nussbaum gain technique (cf. (Su et al., 2009), (Feng, Hong, Chen & Su, 2008), (Feng et al., 2006), (Ge & Wang, 2003)). The above methods are applicable to plants in parametric–pure feedback or controllable form, and with controllers that use the SAB or the MRAC as the control framework.

In (Wang et al., 2004), a system with dead zone in the actuator is considered, assuming that both dead zone slopes have the same value. The input term is rewritten as the sum of an input term with constant control gain plus a bounded disturbance-like term. The disturbance term is rejected by means of a robust technique, based on (Slotine & Li, 1991) pp. 309. Nevertheless, this strategy is not valid for different values of the slopes. Other robustness techniques comprise a control law with compensating terms and either a projection modification of the update law, as in cf. (Huang & Kuo, 2001), or a σ modification as in (Bechlioulis & Rovithakis, 2009), (Li et al., 2004). Nevertheless, some lower or upper bounds of the plant coefficients are required to be known.

The Nussbaum gain technique can relax this requirement, as can be noticed from (Su et al., 2009), (Feng, Hong, Chen & Su, 2008), (Feng, Su & Hong, 2008). The main drawbacks of the Nussbaum gain method are (see (Su et al., 2009), (Feng, Hong, Chen & Su, 2008), (Feng et al., 2006), (Ge & Wang, 2003), (Feng et al., 2007), (Feng, Su & Hong, 2008), (Ren et al., 2008), (Zhang & Ge, 2009), (Du et al., 2010)): i) the upper bound of the transient behavior of the tracking error is significantly modified in comparison with that of the disturbance-free case: the value of this bound depends on the time integral of terms that comprise Nussbaum terms, and ii) the controller involves an additional state, which is necessary to compute the Nussbaum function.

Other drawbacks are: i) the control gain is assumed to be the product of a unknown constant and a known function, as in (Tong et al., 2010), (Liu & Tong, 2010), ii) the control gain is assumed upper bounded by some unknown constant, as in (Zhang & Ge, 2009), (Du et al., 2010), (Su et al., 2009), (Feng, Hong, Chen & Su, 2008), (Feng et al., 2006), (Ge & Wang, 2003), (Feng et al., 2007), (Feng, Su & Hong, 2008), (Ren et al., 2008), iii) the control gain is assumed upper bounded by a known function, as in (Ge & Tee, 2007), (Psillakis, 2010), iv) upper or lower bounds of the plant coefficients are required to be known to achieve asymptotic convergence of the tracking error to a residual of user–defined size, as in (Ge & Tee, 2007), (Chen et al., 2009), (Feng et al., 2006), (Ge & Wang, 2003), (Feng et al., 2007), (Ren et al., 2008), (Ge & Tee, 2007), (Tong et al., 2010), (Liu & Tong, 2010), iv) the control or update law involves signum type signals, as in (Zhang & Ge, 2009), (Du et al., 2010), (Psillakis, 2010), (Su et al., 2009), (Feng, Hong, Chen & Su, 2008), (Feng, Su & Hong, 2008).

Recent adaptive control schemes based on the direct Lyapunov method achieve improved transient performance. For instance, L_1 adaptive control, with the drawback that the control gain is assumed constant, as in (Cao & Hovakimyan, 2006), (Cao & Hovakimyan, 2008a), (Cao & Hovakimyan, 2008b), (Dobrokhodov et al., 2008), (Li & Hovakimyan, 2008).

Other works have the following drawbacks:

i) The control gain is assumed constant, as in (Zhou et al., 2009), (Wen et al., 2009), (Bashash & Jalili, 2009).

ii) The control gain is assumed upper bounded by some unknown constant, as in (Chen, 2009), (Ho et al., 2009) and (Park et al., 2009).

iii) The control gain is assumed upper bounded by some known function as in (Bechlioulis & Rovithakis, 2009).

iv) Upper or lower bounds of plant parameters are required to be known to achieve asymptotic convergence of the tracking error to a residual set of user–defined size, as in (Bashash & Jalili, 2009), (Chen, 2009), (Ho et al., 2009), (Park et al., 2009) and (Bechlioulis & Rovithakis, 2009).

In this chapter, we develop a controller that overcomes the above drawbacks, so that:

Bi) The upper bound of tracking error transient value does not depend on time integral terms.

Bii) Additional states are not used in the controller.

Biii) The control gain is not required to be upper bounded by a constant.

Biv) The control gain is not required to be bounded by a known function.

Bv) Upper or lower bounds of the plant parameters are not required to be known.

Bvi) The control and update laws do not involve signum type signals.

Bvii) The tracking error converges to a residual set whose size is user–defined.

We consider systems described by the controllable form model with arbitrary relative degree, unknown varying but bounded coefficients and varying control gain. We use the SAB of (Kanellakopoulos et al., 1991) as a basic framework for the control design, preserving a simple definition of the states resulting from the backstepping procedure. We use the Lyapunov–like function method to handle the unknown time varying behavior of the plant parameters. All closed loop signals remain bounded so that parameter drifting is prevented.

The key elements to handle the varying behavior of the control gain are: i) introduce the control gain in the term involving the adjusted parameter vector, by means of the inequality that relates the control gain and its lower bound, and ii) apply the Young's inequality.

In current works that deal with plants in controllable form and time varying parameters and use the state transformation based on the backstepping procedure, they modify the defined states at each step of the state transformation in order to tackle the unknown time varying behavior of the plant parameters. Instead of altering the state transformation, we formulate a Lyapunov–like function, such that its magnitude and time derivative vanish when the states resulting from the state transformation reach a target region.

The control design and proof of boundedness and convergence properties are simpler in comparison to current works that use the Nussbaum gain method. The controller is also simpler as it does not introduce additional states that would be necessary to handle the unknown time varying control gain.

The chapter is organized as follows. In section 2 we detail the plant model. In section 3 we present the goal of the control design. In section 4 we carry out a state transformation, based on the state backstepping procedure. In section 5 we derive the control and update laws. In section 6 we prove the boundedness of the closed loop signals. In section 7 we prove the convergence of the tracking error e, finally, in section 8 we present an example.

2. Problem statement

In this section we detail the plant and the reference model. Consider the following plant in controllable form:

$$y^{(n)} = \gamma_n^\top a + bu + d \tag{1}$$

where $y(t) \in \mathbb{R}$ is the system output, $u(t) \in \mathbb{R}$ the input, a a vector of varying entries, γ_n a known vector, b the control gain, and d a disturbance-like term. We make the following assumptions:

Ai) The vector a involves unknown, time varying, bounded entries $a_1, \cdots, a_j$, which satisfy: $|a_1| \leq \bar{\mu}_1, \cdots, |a_j| \leq \bar{\mu}_j$, where $\bar{\mu}_1, \cdots, \bar{\mu}_j$ are unknown, positive constants.

Aii) The entries of the vector γ_n are known linear or nonlinear functions of $y, \cdots, y^{(n-1)}$.

Aiii) The terms y , $\dot{y}, \cdots, y^{(n-1)}$ are available for measurement.

Aiv) The term d represents either external disturbances or unknown model terms that satisfy:

$$|d| \leq \mu_d f_d, \tag{2}$$

where μ_d is an unknown positive constant, and f_d is a known function that depends on $y, \cdots,$ $y^{(n-1)}$. In the case that d is bounded, we have $f_d = 1$. The term d may come from the product of a known function g_d with an unknown varying but bounded coefficient c_g: $d = c_g g_d$, $|c_g| \leq \mu_d$, so that $f_d = |g_d|$, where μ_d is an unknown positive constant whereas g_d is a known function.

Av) The control gain b satisfies:

$$|b| \geq b_m > 0, \; b \neq 0 \; \forall t \geq t_o \tag{3}$$

where b_m is an unknown lower bound, and the value of the signum of b is constant and known.

Remark. We recall that $\mu_d, b_m, \bar{\mu}_1, \cdots, \bar{\mu}_j$ are unknown constants. In contrast, the values of y, $\cdots, y^{(n-1)}, \gamma_n, f_d$, sgn$(b)$ are required to be known. Notice in assumption Av that we do not require the control gain b to be upper bounded by any constant. That is a major contribution with respect to current works that use the Nussbaum gain method, e.g (Su et al., 2009), (Feng, Hong, Chen & Su, 2008), (Feng, Su & Hong, 2008), (Feng et al., 2007), (Ge & Wang, 2003). The requirement about the value of the signum of b is a common and acceptable requirement.

3. Control goal

Let

$$e(t) = y(t) - y_d(t) \tag{4}$$

$$y_d^{(n)} + a_{m,n-1} y_d^{(n-1)} + \cdots + a_{m,o} y_d = a_{m,o} r \tag{5}$$

$$\Omega_e = \{e : |e| \leq C_{be}\} \tag{6}$$

where $e(t)$ is the tracking error, $y_d(t)$ is the desired output, Ω_e is a residual set, r is the reference signal. Moreover, $a_{m,n-1}, \cdots, a_{m,o}$ are constant coefficients defined by the user, such that the polynomial $K(p)$ is Hurwitz, being $K(p)$ defined as $K(p) = p^{(n)} + a_{m,n-1}p^{(n-1)} + \cdots + a_{m,o}$. The reference signal $r(t)$ is bounded and user–defined. The constant C_{be} is positive and user–defined.
The objective of the MRAC design is to formulate a controller, provided by the plant model (1) subject to assumptions Ai to Av, such that:

i) The tracking error e converges asymptotically to the residual set Ω_e.

ii) The control signals are bounded and do not involve discontinuous signals.

4. State transformation based on the state backstepping

In this section we carry out a state transformation by following the steps $0, \cdots, n$ of the backstepping procedure. The plant model (1) can be rewritten as follows:

$$\dot{x}_i = x_{i+1}, \quad 1 \leq i \leq n-1 \tag{7}$$
$$\dot{x}_n = a^\top \gamma_n(x_1, \cdots, x_n) + bu + d \tag{8}$$
$$x_1 = y, \; x_2 = \dot{y}, \; \cdots, \; x_n = y^{(n-1)}$$

The model (7, 8) can be obtained by making $\gamma_1 = \cdots = \gamma_{n-1} = 0$ in the parametric - pure feedback form of (Kanellakopoulos et al., 1991). We use the SAB of (Kanellakopoulos et al., 1991) as the basic framework for the formulation of the control and update laws.

We develop the SAB for the plant model (7, 8), and introduce a new robustness technique. Since the order of the plant is n, the procedure comprises the steps $0, \cdots, n$, to be carried out in a sequential manner.

Step 0. We begin by defining the state z_1 as the tracking error:

$$z_1 = e = y - y_d = x_1 - y_d \tag{9}$$

Step i $(1 \leq i \leq n-1)$. At each i-th step, we obtain the dynamics of the state z_i by deriving it with respect to time, and using the definitions of $\dot{x}_{i+1}$ provided by (7). For the sake of clarity, we develop the step 1 and then we state a generalization for $(1 \leq i \leq n-1)$.

For the case $i = 1$, we differentiate z_1 defined in (9) and use the definition of $\dot{x}_1$ provided by (7) with $i = 1$:

$$\dot{z}_1 = \dot{x}_1 - \dot{y}_d = x_2 - \dot{y}_d = x_2 + \varphi_1 \tag{10}$$
$$\varphi_1 = \varphi_1(\dot{y}_d) = -\dot{y}_d$$

where φ_1 is a known function of $\dot{y}_d$. Equation (10) can be rewritten as:

$$\dot{z}_1 = -c_1 z_1 + z_2 \tag{11}$$
$$z_2 = x_2 + c_1 z_1 - \dot{y}_d \tag{12}$$

where $c_1 \geq 2$ is a positive constant of the user choice. The dynamic equation of z_2 is obtained by differentiating it with respect to time. The same procedure must be followed until the step

$i = n - 1$. To state a generalization, we express $\dot{z}_i$ as:

$$\dot{z}_i = x_{i+1} + \varphi_i, \tag{13}$$

$$\varphi_i = \varphi_i(z_1, \cdots z_i, y_d, \dot{y}_d, \cdots, y_d^{(i)}), \tag{14}$$

where φ_i is a known scalar term, that is function of $z_1, \cdots, z_i, y_d, \dot{y}_d, \cdots, y_d^{(i)}$. Equation (13) can be rewritten as:

$$\dot{z}_i = -c_i z_i + z_{i+1}, \tag{15}$$

$$z_{i+1} = x_{i+1} + \varphi_i + c_i z_i \tag{16}$$

where $c_i \geq 2$ is a positive constant of the user choice. At the step $i = n - 1$ we obtain the dynamic equation for z_{n-1} and the expression for z_n as indicated by (15, 16).

Remark. *Notice that the definition of the states z_i is similar to that of the disturbance free case, so that a simple design is preserved. This is due to the following facts:*

i) Disturbance like terms are absent in the dynamics $\dot{x}_1, \cdots, \dot{x}_{n-1}$ given by (7), so that they are also absent in the dynamics $\dot{z}_1, \cdots, \dot{z}_{n-1}$, as can be noticed in (13).

ii) Dead zone functions of the states z_i are not used.

Step n. We obtain the dynamics of z_n by differentiating it with respect to time and using the expression of $\dot{x}_n$ provided by (8):

$$\dot{z}_n = bu + a^\top \gamma_n + \varphi_n + d \tag{17}$$

$$\varphi_n = \varphi_n(z_1, \cdots, z_n, y_d, \cdots, y_d^{(n)}) \tag{18}$$

where φ_n is a known scalar that is function of $z_1, \cdots, z_n, y_d, \cdots, y_d^{(n)}$. Notice that the disturbance like term d and the control input u appear explicitly at the dynamics of z_n, at the step n of the procedure. Thus, we have completed the state transformation, which allows us to develop the controller.

5. Control and update laws

In this section we develop the control and update laws, taking into account the assumptions stated in section 2 and the goals of section 3. The key elements of the procedure are:

i) Incorporate the assumptions Ai and Aiv, concerning the unknown time varying parameter $a_1, \cdots, a_j$ and the disturbance like term d.

ii) Carry out a linear parameterization.

iii) Express the parameterization in terms of adjustment error and adjusted parameter vector.

iv) Introduce the control gain b within the adjusted parameter vector.

v) Formulate the control law.

vi) Formulate a Lyapunov–like function and find its time derivative.

vii) Formulate the update law.

We begin by rewriting (17) as follows:

$$\dot{z}_n = -c_n z_n + bu + a^\top \gamma_n + \varphi_n + c_n z_n + d, \tag{19}$$

where $c_n \geq 2$ is a positive constant of the user choice. Multiplying (19) by z_n, we obtain:

$$z_n \dot{z}_n = -c_n z_n^2 + b z_n u + z_n a^\top \gamma_n + z_n(\varphi_n + c_n z_n) + z_n d \tag{20}$$

The term $z_n a^\top \gamma_n + z_n(\varphi_n + c_n z_n) + z_n d$ can be rewritten as follows:

$$\begin{aligned} z_n a^\top \gamma_n + z_n(\varphi_n + c_n z_n) + z_n d &= z_n(a^\top \gamma_n + d + \varphi_n + c_n z_n) \\ &= z_n(a_{[1]}\gamma_{n[1]} + \cdots + a_{[j]}\gamma_{n[j]} + d + \varphi_n + c_n z_n) \\ &\leq |z_n| \left(|a_{[1]}\gamma_{n[1]}| + \cdots + |a_{[j]}\gamma_{n[j]}| + |d| + |\varphi_n + c_n z_n| \right) \end{aligned} \tag{21}$$

using assumptions Ai and Aiv of section 2 and parameterizing, we obtain:

$$\begin{aligned} & z_n a^\top \gamma_n + z_n(\varphi_n + c_n z_n) + z_n d \\ &\quad \leq |z_n| \left(\bar{\mu}_1 |\gamma_{n[1]}| + \cdots + \bar{\mu}_j |\gamma_{n[j]}| + \mu_d f_d + |\varphi_n + c_n z_n| \right) \\ &\quad = \sqrt{b_{mn}} |z_n| \bar{\varphi}^\top \theta \end{aligned} \tag{22}$$

where

$$\bar{\varphi} = \left[|\gamma_{n[1]}|, \ \cdots, \ |\gamma_{n[j]}|, \ f_d, \ |\varphi_n + c_n z_n| \right]^\top \tag{23}$$

$$\theta = (1/\sqrt{b_{mn}}) \left[\bar{\mu}_1, \ \cdots, \ \bar{\mu}_j, \mu_d, \ 1 \right]^\top \tag{24}$$

Notice that the entries of the vector θ are unknown, positive, constant, because bounds of the time varying parameters a_i and d have been introduced, according to the properties in assumptions Ai and Aiv of section 2. Now, we express (22) in terms of adjustment error and adjusted parameter vector:

$$z_n a^\top \gamma_n + z_n(\varphi_n + c_n z_n) + z_n d \leq -\sqrt{b_{mn}} |z_n| \bar{\varphi}^\top \tilde{\theta} + \sqrt{b_{mn}} |z_n| \bar{\varphi}^\top \hat{\theta} \tag{25}$$

where

$$\begin{aligned} \tilde{\theta} &= \hat{\theta} - \theta \\ &= \hat{\theta} - \frac{1}{\sqrt{b_{mn}}} \left[\bar{\mu}_1, \ \cdots, \ \bar{\mu}_j, \ \mu_d, \ 1 \right]^\top \end{aligned} \tag{26}$$

being $\hat{\theta}$ an adjusted parameter vector and $\tilde{\theta}$ an adjustment error. Using the property (3) in the term $\sqrt{b_{mn}} |z_n| \bar{\varphi}^\top \hat{\theta}$ of (25), we obtain:

$$\sqrt{b_{mn}} |z_n| |\bar{\varphi}^\top \hat{\theta}| \leq \sqrt{\frac{3C_{bvz}}{2}} \sqrt{\frac{2}{3C_{bvz}}} \sqrt{|b|} |z_n| |\bar{\varphi}^\top \hat{\theta}| \tag{27}$$

$$\text{where } C_{bvz} = (1/2) C_{be}^2 \tag{28}$$

using the Young's inequality (cf. (Royden, 1988) pp. 123), we obtain:

$$\sqrt{b_{mn}}|z_n||\bar{\varphi}^\top\hat{\theta}| \leq \frac{3}{4}C_{bvz} + \frac{1}{3C_{bvz}}|b|z_n^2(\bar{\varphi}^\top\hat{\theta})^2 \tag{29}$$

Substituting (29) into (25), we obtain:

$$\begin{aligned} & z_n a^\top \gamma_n + z_n(\varphi_n + c_n z_n) + z_n d \\ & \leq -\sqrt{b_{mn}}|z_n|\bar{\varphi}^\top\tilde{\theta} + \tfrac{3}{4}C_{bvz} + \tfrac{1}{3C_{bvz}}|b|z_n^2(\bar{\varphi}^\top\hat{\theta})^2 \end{aligned} \tag{30}$$

Remark. *We have proposed a new method to handle the unknown varying behavior of the control gain b, alternative to the current Nussbaum gain method. We parameterized the model term $z_n a^\top \gamma_n + z_n(\varphi_n + c_n z_n) + z_n d$ in terms of adjustment error $\tilde{\theta}$ and adjusted parameter vector $\hat{\theta}$, and developed the following steps:*

i) Introduce the constant $\sqrt{b_{mn}}$ in the parameterization, see (22).

ii) Introduce the inequality $\sqrt{b_{mn}} \leq \sqrt{|b|}$, see (27).

iii) Apply the Young's inequality to obtain b, see (29).

Recall that the value of b_{mn} is not required to be known.

Substituting (30) into (20), we obtain:

$$\begin{aligned} & z_n\dot{z}_n \leq -c_n z_n^2 + (3/4)C_{bvz} + b z_n\left(u + \tfrac{1}{3C_{bvz}}\operatorname{sgn}(b) z_n(\bar{\varphi}^\top\hat{\theta})^2\right) \\ & -\sqrt{b_{mn}}|z_n|\bar{\varphi}^\top\tilde{\theta} \end{aligned} \tag{31}$$

we choose the following control law:

$$u = -\frac{1}{3C_{bvz}}\operatorname{sgn}(b) z_n(\bar{\varphi}^\top\hat{\theta})^2 \tag{32}$$

where $\bar{\varphi}$, z_n are defined in (23), (16), respectively. Substituting (32) into (31), we obtain:

$$z_n\dot{z}_n \leq -c_n z_n^2 + (3/4)C_{bvz} - \sqrt{b_{mn}}|z_n|\bar{\varphi}^\top\tilde{\theta} \tag{33}$$

To handle the effect of the constant $(3/4)C_{bvz}$, we formulate the following Lyapunov–like function:

$$\bar{V}_z = \begin{cases} (1/2)(\sqrt{V_z} - \sqrt{C_{bvz}})^2 & \text{if } V_z \geq C_{bvz} \\ 0 & \text{otherwise} \end{cases} \tag{34}$$

$$V_z = (1/2)(z_1^2 + \cdots + z_n^2) \tag{35}$$

where C_{bvz} is defined in (28). We need the following properties:

Proposition 5.1. *The function $\bar{V}_z$ defined in* (34) *has the following properties:*

$$i)\bar{V}_z \geq 0 \tag{36}$$

$$ii)V_z \leq 3C_{bvz} + 3\bar{V}_z \tag{37}$$

$$iii)\bar{V}_z,\ (\partial\bar{V}_z/\partial V_z) \text{ are continuous with respect to } V_z \tag{38}$$

Proof. From (34) it follows that $\bar{V}_z \geq 0 \forall t \geq t_o$, the property i of proposition 5.1. In addition, from (34) it follows that

$$V_z \leq (\sqrt{2\bar{V}_z} + \sqrt{C_{bvz}})^2 \tag{39}$$

Applying the Young's inequality (cf. (Royden, 1988) pp. 123), we obtain:

$$V_z = C_{bvz} + 2\sqrt{C_{bvz}}\sqrt{2\bar{V}_z} + 2\bar{V}_z \leq 3C_{bvz} + 3\bar{V}_z \tag{40}$$

This completes the proof of property ii. From (42) it follows that $\partial\bar{V}_z/\partial V_z = 0$ if $V_z = C_{bvz}$ and that $\partial\bar{V}_z/\partial V_z$ is continuous. From (34) it follows that $\bar{V}_z$ is continuous. This completes the proof of property iii of proposition 5.1.

□

Differentiating (34) with respect to time, we obtain:

$$\frac{d\bar{V}_z}{dt} = \frac{\partial\bar{V}_z}{\partial V_z}\dot{V}_z \tag{41}$$

$$\frac{\partial\bar{V}_z}{\partial V_z} = \begin{cases} (1/2)(1/\sqrt{V_z})(\sqrt{V_z} - \sqrt{C_{bvz}}) & \text{if } V_z \geq C_{bvz} \\ 0 & \text{otherwise} \end{cases} \tag{42}$$

To compute $\dot{V}_z$, we differentiate (35) with respect to time: $\dot{V}_z = z_1\dot{z}_1 + \cdots + z_n\dot{z}_n$. Introducing (11) and (15), we obtain

$$\dot{V}_z = -c_1 z_1^2 + z_1 z_2 - c_2 z_2^2 + z_2 z_3 + \cdots + z_n \dot{z}_n \tag{43}$$

substituting (33), we obtain:

$$\begin{aligned}\dot{V}_z \leq & -c_1 z_1^2 + z_1 z_2 - c_2 z_2^2 + z_2 z_3 + \cdots - c_n z_n^2 \\ & +(3/4)C_{bvz} - \sqrt{b_{mn}}|z_n|\bar{\varphi}^\top\tilde{\theta}\end{aligned} \tag{44}$$

Provided that $c_1 \geq 2, c_2 \geq 2, \cdots, c_n \geq 2$ and completing the squares yields:

$$\begin{aligned}& -c_1 z_1^2 + z_1 z_2 - c_2 z_2^2 + z_2 z_3 \cdots - c_n z_n^2 \\ & \leq -z_1^2 - (3/4)z_2^2 + \cdots - (3/4)z_n^2 \leq -(3/2)V_z\end{aligned}$$

substituting into (44), we obtain:

$$\dot{V}_z \leq -(3/2)V_z + (3/4)C_{bvz} - \sqrt{b_{mn}}|z_n|\bar{\varphi}^\top\tilde{\theta} \tag{45}$$

Since $\partial\bar{V}_z/\partial V_z$ is non-negative, we can multiply it into (45) without changing the direction of the inequality:

$$\frac{\partial\bar{V}_z}{\partial V_z}\dot{V}_z \leq -(3/2)V_z\frac{\partial\bar{V}_z}{\partial V_z} + (3/4)C_{bvz}\frac{\partial\bar{V}_z}{\partial V_z} - \sqrt{b_{mn}}|z_n|\bar{\varphi}^{\top}\tilde{\theta}\frac{\partial\bar{V}_z}{\partial V_z} \tag{46}$$

Substituting into (41), we obtain:

$$\frac{d\bar{V}_z}{dt} \leq -(3/2)V_z\frac{\partial\bar{V}_z}{\partial V_z} + (3/4)C_{bvz}\frac{\partial\bar{V}_z}{\partial V_z} - \sqrt{b_{mn}}|z_n|\bar{\varphi}^{\top}\tilde{\theta}\frac{\partial\bar{V}_z}{\partial V_z} \tag{47}$$

we choose the update law so as to reject the effect of the term involving the adjustment error $\tilde{\theta}$:

$$\dot{\hat{\theta}} = \Gamma\bar{\varphi}|z_n|\frac{\partial\bar{V}_z}{\partial V_z} \tag{48}$$

where Γ is a diagonal matrix whose diagonal elements are positive constants defined by the user, whereas $\bar{\varphi}$, z_n, $\partial\bar{V}_z/\partial V_z$ are defined in (23), (16), (42), respectively.

Remark. *So far, we have developed the controller, which involves the control law* (32) *and the update law* (48). *Other parameters necessary for its implementation are: V_z defined in* (35)*; $z_1, z_2, \cdots, z_n$ defined in* (9), (12), $\cdots$, (16), *respectively; C_{bvz} defined in* (28). *Recall that $c_1 \geq 2, \cdots, c_n \geq 2$ are user–defined positive constants.*

Remark. *The control and update laws stated in* (32) *and* (48) *have the following features:*

i) The control law uses the adjusted parameter vector $\hat{\theta}$, so that it does not rely on upper or lower bounds of the plant coefficients, i.e. $\bar{\mu}_1, \cdots, \bar{\mu}_j, \mu_d, b_{mn}$, and excessive control effort is also avoided.

ii) Additional states are not required to handle the unknown varying behavior of the control gain, what is an important benefit with respect to closely related schemes that use the Nussbaum gain method.

iii) The control and update laws do not involve discontinuous signals. In fact, the vectorial field of the closed loop system is Lipschitz continuous, so that trajectory unicity is preserved.

6. Boundedness analysis

In this section we prove that the closed loop signals $z_1, \cdots, z_n, \hat{\theta}, u$ are bounded if the developed controller is applied.

Theorem 6.1. Boundedness of the closed loop signals. *Consider the plant* (1) *subject to assumptions Ai to Av; the signals $z_1, \cdots, z_n$ defined in* (9), (12) *and* (16)*; the signals $\bar{\varphi}$, V_z, $\partial\bar{V}_z/\partial V_z$, C_{bvz} defined in* (23), (35), (42) *and* (28), *respectively. If the controller* (32), (48) *is applied, then the signals $z_1, \cdots, z_n, \hat{\theta}$, and u remain bounded and $|e|$ is upper bounded as follows:*

$$|e| \leq \sqrt{2}\left(\sqrt{C_{bvz}} + \sqrt{2V(\bar{x}(t_o))}\right)^2 \tag{49}$$

Proof. We choose the following Lyapunov–like function:

$$V(\bar{x}(t)) = \bar{V}_z + V_\theta \tag{50}$$

$$V_\theta = (1/2)\sqrt{b_{mn}}\tilde{\theta}^\top\Gamma^{-1}\tilde{\theta} \tag{51}$$
$$\bar{x}(t) = [z_1, \ \ldots, \ z_n, \ \tilde{\theta}^\top]^\top \tag{52}$$

where $\bar{V}_z$ is defined in (34) and $\tilde{\theta}$ in (26). The time derivative of V_θ is:

$$\dot{V}_\theta = (1/2)\sqrt{b_{mn}}(\dot{\tilde{\theta}}^\top\Gamma^{-1}\tilde{\theta} + \tilde{\theta}^\top\Gamma^{-1}\dot{\tilde{\theta}}) \tag{53}$$

Since Γ is diagonal, then Γ^{-1} is diagonal, $(\Gamma^{-1})^\top = \Gamma^{-1}$, $\dot{\tilde{\theta}}^\top\Gamma^{-1}\tilde{\theta} = \tilde{\theta}^\top\Gamma^{-1}\dot{\tilde{\theta}}$. In view of this and the update law (48), we have:

$$\dot{V}_\theta = \sqrt{b_{mn}}\tilde{\theta}^\top\Gamma^{-1}\dot{\hat{\theta}} = \sqrt{b_{mn}}\tilde{\theta}^\top\bar{\varphi}|z_n|\frac{\partial\bar{V}_z}{\partial V_z} \tag{54}$$

Differentiating (50) with respect to time, we obtain: $\dot{V} = \dot{\bar{V}}_z + \dot{V}_\theta$. Substituting equations (47) and (54), we obtain:

$$\begin{aligned}\dot{V} &\leq -(3/2)V_z\tfrac{\partial\bar{V}_z}{\partial V_z} + (3/4)C_{bvz}\tfrac{\partial\bar{V}_z}{\partial V_z}\\ &= -\tfrac{3}{2}\tfrac{\partial\bar{V}_z}{\partial V_z}\left(\tfrac{V_z}{2} + \tfrac{V_z}{2} - \tfrac{C_{bvz}}{2}\right)\end{aligned} \tag{55}$$

From (42) if follows that

$$\partial\bar{V}_z/\partial V_z = 0 \text{ for } V_z \leq C_{bvz} \tag{56}$$
$$\partial\bar{V}_z/\partial V_z > 0 \text{ for } V_z > C_{bvz}. \tag{57}$$

In view of this and (55), we obtain:

$$\begin{aligned}&\dot{V} \leq -\frac{3}{2}\frac{\partial\bar{V}_z}{\partial V_z}\left(\frac{V_z}{2}\right) \text{ if } V_z \geq C_{bvz}\\ &\dot{V} \leq 0 = -\frac{3}{2}\frac{\partial\bar{V}_z}{\partial V_z}\left(\frac{V_z}{2}\right) \text{ if } V_z < C_{bvz}\\ &\Rightarrow \dot{V} \leq -\frac{3}{4}\frac{\partial\bar{V}_z}{\partial V_z}V_z \leq 0\end{aligned} \tag{58}$$

Thus, $\dot{V} + 0V \leq 0$. Using the Lemma in (Slotine & Li, 1991) pp. 91, we obtain:

$$V(\bar{x}(t)) \leq V(\bar{x}(t_o))\exp(-0t) = V(\bar{x}(t_o)) \tag{59}$$

where

$$V(\bar{x}(t_o)) = \bar{V}_{zo} + V_{\theta o} \tag{60}$$
$$\bar{V}_{zo} = \begin{cases}(1/2)(\sqrt{V_{zo}} - \sqrt{C_{bvz}})^2 \text{ if } V_{zo} \geq C_{bvz}\\ 0 \qquad \text{otherwise}\end{cases} \tag{61}$$
$$V_{zo} = (1/2)(z_1(t_o)^2 + \cdots + z_n(t_o)^2) \tag{62}$$
$$V_{\theta o} = (1/2)\sqrt{b_{mn}}(\hat{\theta}(t_o) - \theta)^\top\Gamma^{-1}(\hat{\theta}(t_o) - \theta) \tag{63}$$

Since $V(\bar{x}(t)) \geq 0$, we have: $0 \leq V(\bar{x}(t)) \leq V(\bar{x}(t_o))$. Introducing the definition (50), we obtain

$$\bar{V}_z + V_\theta \leq V(\bar{x}(t_o)) \tag{64}$$

$$\Rightarrow \bar{V}_z \leq V(\bar{x}(t_o)),\ V_\theta \leq V(\bar{x}(t_o)) \tag{65}$$

Thus, it follows from (51) that $\tilde{\theta} \in L_\infty$, and consequently $\hat{\theta} \in L_\infty$. The inequality $\bar{V}_z \leq V(\bar{x}(t_o))$ implies that the tracking error e is bounded, as we show hereafter. We begin by solving (34) for V_z:

$$V_z = \left(\sqrt{C_{bvz}} + \sqrt{2\bar{V}_z}\right)^2 \text{ if } \bar{V}_z > 0 \tag{66}$$

$$V_z \leq C_{bvz} \text{ if } \bar{V}_z = 0 \tag{67}$$

Using the inequality $\bar{V}_z \leq V(\bar{x}(t_o))$, we obtain:

$$V_z \leq \left(\sqrt{C_{bvz}} + \sqrt{2V(\bar{x}(t_o))}\right)^2 \text{ if } \bar{V}_z > 0 \tag{68}$$

$$V_z \leq C_{bvz} \leq \left(\sqrt{C_{bvz}} + \sqrt{2V(\bar{x}(t_o))}\right)^2 \text{ if } \bar{V}_z = 0 \tag{69}$$

combining both inequalities, we obtain:

$$V_z \leq \left(\sqrt{C_{bvz}} + \sqrt{2V(\bar{x}(t_o))}\right)^2 \tag{70}$$

Introducing the definition (35), we obtain:

$$\sqrt{z_1^2 + \cdots + z_n^2} \leq \sqrt{2}\left(\sqrt{C_{bvz}} + \sqrt{2V(\bar{x}(t_o))}\right)^2 \tag{71}$$

so that $z_1 \in L_\infty, \cdots, z_n \in L_\infty$. Since $e^2 = z_1^2 \leq z_1^2 + \cdots + z_n^2$, we obtain:

$$|e| \leq \sqrt{2}\left(\sqrt{C_{bvz}} + \sqrt{2V(\bar{x}(t_o))}\right)^2 \tag{72}$$

which indicates the upper bound for the tracking error e.

Remark. *Notice that this upper bound does not involve integral terms, what is an important advantage with respect to the Nussbaum Gain method, see (Su et al., 2009), (Feng, Hong, Chen & Su, 2008), (Feng et al., 2006), (Ge & Wang, 2003), (Feng et al., 2007), (Feng, Su & Hong, 2008), (Ren et al., 2008).*

Now, we proceed to show the boundedness of u. From (9), (12), (16) it follows that $x_1 \in L_\infty$, $x_2 \in L_\infty, \cdots, x_n \in L_\infty$. Therefore, $\gamma_n \in L_\infty$. It follows from (23) that $\bar{\varphi} \in L_\infty$. From (32) it follows that $u \in L_\infty$. This completes the proof.

□

7. Convergence analysis

In this section we prove that if the developed controller is applied, then the signal z_1 converges asymptotically to Ω_z, where $\Omega_z = \{z_1 : |z_1| \leq C_{be}\}$.

Theorem 7.1. Convergence of the tracking error. *Consider the plant* (1) *subject to assumptions Ai to Av; the signals* $z_1, \cdots, z_n$ *defined in* (9), (12) *and* (16)*; the signals* $\bar{\varphi}$, V_z, $\partial\bar{V}_z/\partial V_z$, C_{bvz} *defined in* (23), (35), (42) *and* (28)*, respectively. If the controller* (32), (48) *is applied, then the signal* z_1 *converges asymptotically to* Ω_z*, where* $\Omega_z = \{z_1 : |z_1| \leq C_{be}\}$.

Proof. In view of (42), equation (58) can be rewritten as:

$$\dot{V} \leq -f_d \leq 0 \tag{73}$$

$$f_d = \begin{cases} (3/8)(\sqrt{V_z})(\sqrt{V_z} - \sqrt{C_{bvz}}) & \text{if } V_z \geq C_{bvz} \\ 0 & \text{otherwise} \end{cases} \tag{74}$$

The derivative $\partial f_d/\partial V_z$ is not continuous, as it involves an abrupt change at $V = C_{bvz}$. Thus, the Barbalat's Lemma can not be applied on f_d. To remedy that, we shall express (73) in terms of a function with continuous derivative:

$$\dot{V} \leq -f_d \leq -f_g \leq 0 \tag{75}$$

$$f_g = \begin{cases} (3/8)(\sqrt{V_z} - \sqrt{C_{bvz}})^2 & \text{if } V_z \geq C_{bvz} \\ 0 & \text{otherwise} \end{cases} \tag{76}$$

Arranging and integrating (75), we obtain:

$$f_g \leq -\dot{V}$$

$$\int_{to}^{t} f_g d\tau \leq V(\bar{x}(t_o)) - V(\bar{x}(t))$$

$$V(\bar{x}(t)) + \int_{to}^{t} f_g d\tau \leq V(\bar{x}(t_o)) \tag{77}$$

Thus, $f_g \in L_1$. We have to prove that $f_g \in L_\infty$, $\dot{f}_g \in L_\infty$ to apply the Barbalat's Lemma. Since $V_z \in L_\infty$, it follows from (76) that $f_g \in L_\infty$. Differentiating (76) with respect to time, we obtain:

$$\dot{f}_g = \frac{\partial f_g}{\partial V_z}\dot{V}_z \tag{78}$$

$$\frac{\partial f_g}{\partial V_z} = \begin{cases} (3/8)(1/\sqrt{V_z})(\sqrt{V_z} - \sqrt{C_{bvz}}) & \text{if } V_z \geq C_{bvz} \\ 0 & \text{otherwise} \end{cases} \tag{79}$$

Notice that $\partial f_g/\partial V_z$ is continuous. Since $V_z \in L_\infty$, then $\partial f_g/\partial V_z \in L_\infty$. Since $z_1 \in L_\infty, \cdots$, $z_n \in L_\infty$, it follows from (11), (15) that $\dot{z}_1 \in L_\infty, \cdots, \dot{z}_{n-1} \in L_\infty$. Since $u \in L_\infty$, it follows from (17) that $\dot{z}_n \in L_\infty$. Therefore, from (43) it follows that $\dot{V}_z \in L_\infty$.

So far we have proved that $\partial f_g/\partial V_z \in L_\infty$ and $\dot{V}_z \in L_\infty$, so that it follows from (78) that $\dot{f}_g \in L_\infty$. In view of $f_g \in L_\infty$, $\dot{f}_g \in L_\infty$, application of Barbalat's Lemma (cf. (Ioannou & Sun, 1996) pp. 76), then indicates that f_g converges asymptotically to zero. Hence, from (76) it follows that V_z converges to Ω_{vz}, where $\Omega_{vz} = \{V_z : V_z \leq C_{bvz}\}$. From the definition (35), it follows that z_1 converges asymptotically to Ω_z, where $\Omega_z = \{z_1 : |z_1| \leq \sqrt{2C_{bvz}}\}$. Since $C_{bvz} = (1/2)C_{be}^2$, it follows that $\Omega_z = \{z_1 : |z_1| \leq C_{be}\}$. This completes the proof. □

8. Simulation example

Consider the following case of the plant (1):

$$\ddot{y} = \gamma_2^\top a + bu + d \tag{80}$$

$$\gamma_2 = [\dot{y},\ y]^\top,\ a = [a_1,\ a_2]^\top \tag{81}$$

$$\begin{aligned} a_1 &= -2\,(1+0.1\sin(2(\pi/8)t))\,,\ a_2 = -1\,(1+0.1\sin(2(\pi/5)t)) \\ b &= 2\,(1+0.1\sin((2\pi/11)t)) + 0.6|y| \\ d &= -0.2\,(1+0.1\sin((2\pi/7)t))\,y \end{aligned} \tag{82}$$

The aim is that y converges towards y_d, with a threshold of 0.1. In figure 1 we present a simulation block diagram for the example.

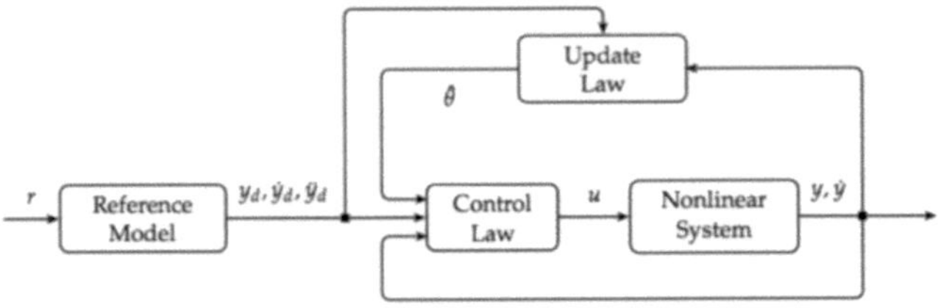

Fig. 1. Simulation block diagram.

The properties Ai, Aiv, Av of section 2 are analyzed at the following. From (82) it follows that a_1, d, b are bounded as:

$$|a_1| \leq 2(1.1) = 2.2,\ \ |a_2| \leq 1(1.1) = 1.1 \tag{83}$$

$$|d| \leq 0.2(1.1)|y| = 0.22|y| \tag{84}$$

$$|b| \geq 2(0.9) = 1.8 > 0 \tag{85}$$

Hence, the upper bounds of a_1, a_2, d, and the lower bound of b, are:

$$|a_1| \leq \bar{\mu}_1 = 2.2,\ |a_2| \leq \bar{\mu}_2 = 1.1,\ \ |b| \geq b_{mn} = 1.8 \tag{86}$$

$$|d| \leq \mu_d f_d,\ \mu_d = 0.22,\ f_d = |y| \tag{87}$$

where f_d is not constant and known, whereas $b_{mn}, \bar{\mu}_1, \bar{\mu}_2, \mu_d, b_{mn}$ are positive, constant and unknown to the controller. From (86), (87) it follows that assumptions Ai, Aiv, Av of section 2 are satisfied.

The procedure of section 4 is followed in order to establish the terms involved in the control and update laws, mentioned in remark 5. Eq. (80) can be rewritten as:

$$\dot{x}_1 = x_2 \tag{88}$$

$$\dot{x}_2 = \gamma_2^\top a + bu + d \tag{89}$$

$$x_1 = y,\ x_2 = \dot{y},\ n = 2 \tag{90}$$

since $n = 2$, the state transformation based on the backstepping procedure involves the steps $0, 1, 2$.

Step 0. Let

$$z_1 = e = y - y_d = x_1 - y_d \tag{91}$$

as in (9).

Step 1. Differentiating (91) with respect to time and arranging, yields:

$$\dot{z}_1 = -c_1 z_1 + z_2 \tag{92}$$

$$z_2 = x_2 + c_1 z_1 - \dot{y}_d \tag{93}$$

as in (11), (12).

Step 2. Since $n = 2$, the second step is the last one. Differentiating (93) with respect to time, using (89) and arranging, yields:

$$\begin{aligned}\dot{z}_2 &= \dot{x}_2 + c_1 \dot{z}_1 - \ddot{y}_d \\ &= \gamma_2^\top a + bu + d + c_1 \dot{z}_1 - \ddot{y}_d \\ &= \gamma_2^\top a + bu + d + c_1(x_2 + \varphi_1) - \ddot{y}_d\end{aligned} \tag{94}$$

using the definitions (91), (93), yields:

$$\dot{z}_2 = \gamma_2^\top a + bu + d + \varphi_2 \tag{95}$$

$$\varphi_2 = c_1(z_2 - c_1 z_1) - \ddot{y}_d \tag{96}$$

notice that the form of (95), (96) is that of (17), (18), respectively. This completes the state transformation based on the backstepping procedure.

The parameters defined above can be summarized as:

$$z_1 = y - y_d \tag{97}$$

$$z_2 = x_2 + c_1 z_1 - \dot{y}_d \tag{98}$$

$$x_1 = y, \ x_2 = \dot{y}_d \tag{99}$$

$$\varphi_1 = -\dot{y}_d \tag{100}$$

$$\varphi_2 = c_1(z_2 - c_1 z_1) - \ddot{y}_d \tag{101}$$

According to remark 5, it remains to define $\bar{\varphi}$, V_z. From (81), definition (23) and $n = 2$, it follows that

$$\bar{\varphi} = \left[|\gamma_{2[1]}|, \ |\gamma_{2[2]}|, \ f_d, \ |\varphi_2 + c_2 z_2|\right]^\top = [|\dot{y}|, \ |y|, \ f_d, \ |\varphi_2 + c_2 z_2|]^\top \tag{102}$$

From (35) and $n = 2$ it follows that

$$V_z = (1/2)(z_1^2 + z_2^2) \tag{103}$$

Expressions (97) to (103) allow to define the control and update law. From (32), (48), (82) and $n = 2$ it follows that

$$\operatorname{sgn}(b) = +1 \tag{104}$$

$$u = -\frac{1}{3C_{bvz}} z_2(\bar{\varphi}^\top \hat{\theta})^2 \tag{105}$$

$$\dot{\hat{\theta}} = \Gamma \bar{\varphi} |z_2| \frac{\partial \bar{V}_z}{\partial V_z} \tag{106}$$

the main parameters needed to compute u and $\hat{\theta}$ are: $\bar{\varphi}$ (102), φ_2 (101), C_{bvz} (28), z_2 (98), z_1 (97), $\partial \bar{V}_z / \partial V_z$ (42), V_z (103). In addition, Γ is a diagonal matrix whose diagonal elements are positive constants defined by the user.

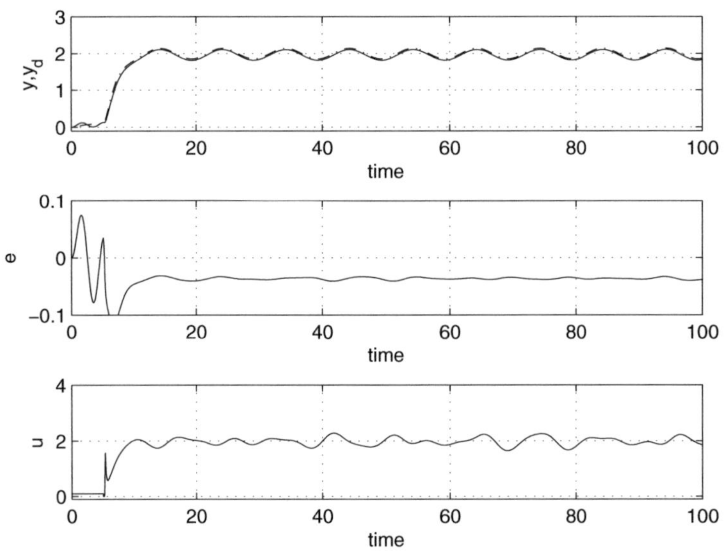

Fig. 2. Example 1, upper: output y (continuous line), desired output y_d (dash–dot line); middle: tracking error e; lower: control input u.

Since the aim is that y converges towards y_d, with a threshold of 0.1, we set $C_{be} = 0.1$. We use the reference model (5) with $y_d(t_o) = y(t_o)$, $\dot{y}_d(t_o) = 0$, $a_{m,1} = 1$, $a_{m,o} = 1$. We use the following parameter values for the control and update laws: $c_1 = 2$, $c_2 = 2$, $\Gamma = \text{diag}\{1,\ 1,\ 1,\ 1\}$.

The results are shown in figures 2 and 3. We have choosen $y_d(t_0) \approx y(t_0)$ in order to obtain a rapid convergence of y towards y_d. Figure 2 shows that. *i)* the tracking error e converges asymptotically towards $\Omega_e = \{e : |e| \leq 0.1\}$. *ii)* The output y converges towards y_d with threshold 0.1 without large transient differences. Figure 3 shows that $\hat{\theta}_1, ..., \hat{\theta}_4$ are not decreasing with respect to time. This occurs because $\dot{\hat{\theta}}$ is non-negative. The procedure for the sample plant (80) is simpler in comparison with adapive controllers that use the Nussbaum gain method.

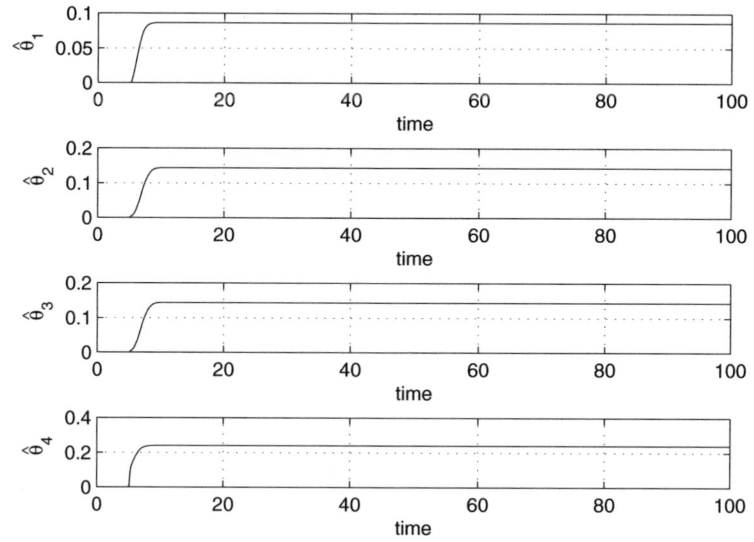

Fig. 3. Example 1, entries of the updated parameter vector $\hat{\theta}$, from upper to lower: $\hat{\theta}_1$; $\hat{\theta}_2$, $\hat{\theta}_3$, $\hat{\theta}_4$.

9. Acknowledgements

A. Rincon acknowledges financial support provided by "Programa de becas para estudiantes sobresalientes de posgrado", Universidad Nacional de Colombia - vicerrectoría de Investigación. This work was partially supported by Universidad Nacional de Colombia - Manizales, project 12475, Vicerrectoría de Investigación, DIMA.

10. References

Bashash, S. & Jalili, N. (2009). Robust adaptive control of coupled parallel piezo–flexural nanopositioning stages, *IEEE/ASME Transactions on Mechatronics* Vol. 14(No. 1): 11–20.

Bechlioulis, C. & Rovithakis, G. (2009). Adaptive control with guaranteed transient and steady state tracking error bounds for strict feedback systems, *Automatica* Vol. 45(No. 2): 532–538.

Cao, C. & Hovakimyan, N. (2006). Design and analysis of a novel l_1 adaptive controller, part i: control signal and asymptotic stability, *Proceedings of the 2006 American Control Conference*, Publisher, Minneapolis (USA), pp. 3397–3402.

Cao, C. & Hovakimyan, N. (2008a). Design and analysis of a novel adaptive control architecture with guaranteed transient performance, *IEEE Transactions on Automatic Control* Vol. 53(No. 2): 586–591.

Cao, C. & Hovakimyan, N. (2008b). l_1 adaptive output feedback controller for systems of unknown dimension, *IEEE Transactions on Automatic Control* Vol. 53(No. 3): 815–821.

Chen, B., Liu, X., Liu, K. & Lin, C. (2009). Novel adaptive neural control design for nonlinear mimo time–delay systems, *Automatica* Vol. 45(No. 6): 1554–1560.

Chen, C., Lin, C. & Chen, T. (2008). Intelligent adaptive control for mimo uncertain nonlinear systems, *Expert Systems with Applications* Vol. 35(No. 3): 865 – 877.

Chen, W. (2009). Adaptive backstepping dynamic surface control for systems with periodic disturbances using neural networks, *IET Control Theory and Applications* Vol. 3(No. 10): 1383–1394.

Dobrokhodov, V., Kitsios, I., Kaminer, I. & Jones, K. (2008). Flight validation of a metrics driven l_1 adaptive control, *Proceedings of AIAA Guidance, Navigation and Control Conference and Exhibit*, Publisher, Honolulu (USA), pp. 1–22.

Du, H., Ge, S. & Liu, J. (2010). Adaptive neural network output feedback control for a class of non–afine non–linear systems with unmodelled dynamics, *IET Control Theory and Applications* Vol. 5(No. 3): 465–477.

Feng, Y., Hong, H., Chen, X. & Su, C. (2008). Robust adaptive controller design for a class of nonlinear systems preceded by generalized prandtl-ishlinskii hysteresis representation, *Proceedings of 7th World Congress on Intelligent Control and Automation*, Publisher, Chongqing (China), pp. 382–387.

Feng, Y., Hu, Y. & Su, C. (2006). Robust adaptive control for a class of perturbed strict-feedback nonlinear systems with unknown prandtl-ishlinskii hysteresis, *Proceedings of the 2006 IEEE International Symposium on Intelligent Control*, Publisher, Munich (Germany), pp. 106–111.

Feng, Y., Su, C. & Hong, H. (2008). Universal contruction of robust adaptive control laws for a class of nonlinear systems preceded by generalized prandtl-ishlinskii representation, *Proceedings of the 3rd IEEE Conference on Industrial Electronics and Applications*, Publisher, Singapore, pp. 153–158.

Feng, Y., Su, C., Hong, H. & Ge, S. (2007). Robust adaptive control for a class of nonlinear systems with generalized prandtl-ishlinskii hysteresis, *Proceedings of the 46th IEEE Conference on Decision and Control*, Publisher, New Orleans (LA, USA), pp. 4833–4838.

Ge, S. & Tee, K. (2007). Approximation–based control of nonlinear mimo time–delay systems, *Automatica* Vol. 43(No. 1): 31–43.

Ge, S. & Wang, J. (2003). Robust adaptive tracking for time-varying uncertain nonlinear systems with unknown control coefficients, *IEEE Transactions on Automatic Control* Vol. 48(No. 8): 1463–1469.

Ho, H., Wong, Y. & Rad, A. (2009). Adaptive fuzzy sliding mode control with chattering elimination for nonlinear siso systems, *Simulation Modelling Practice and Theory* Vol. 17(No. 7): 1199–1210.

Hong, Y. & Yao, B. (2007). A globally stable saturated desired compensation adaptive robust control for linear motor systems with comparative experiments, *Automatica* Vol. 43(No. 10): 1840–1848.

Hsu, C., Lin, C. & Lee, T. (2006). Wavelet adaptive backstepping control for a class of nonlinear systems, *IEEE Transactions on Neural Networks* Vol. 17(No. 5): 1175–1183.

Huang, A. & Kuo, Y. (2001). Sliding control of non-linear systems containing time-varying uncertainties with unknown bounds, *International Journal of Control* Vol. 74(No. 3): 252–264.

Ioannou, P. & Sun, J. (1996). *Robust Adaptive Control*, Prentice-Hall PTR Upper Saddle River, New Jersey.

Jiang, Z. & Hill, D. (1999). A robust adaptive backstepping scheme for nonlinear systems with unmodeled dynamics, *IEEE Transactions on Automatic Control* Vol. 44(No. 9): 1705–1711.

Kanellakopoulos, I., Kokotović, P. & Morse, A. (1991). Systematic design of adaptive controllers for feeedback linearizable systems, *IEEE Transactions on Automatic Control* Vol. 36(No. 11): 1241 – 1253.

Koo, K. (2001). Stable adaptive fuzzy controller with time-varying dead zone, *Fuzzy Sets and Systems* Vol. 121(No. 1): 161–168.

Labiod, S. & Guerra, T. (2007). Adaptive fuzzy control of a class of siso nonaffine nonlinear systems, *Fuzzy Sets and Systems* Vol. 158(No. 10): 1126 – 1137.

Li, D. & Hovakimyan, N. (2008). Filter design for feedback–loop trade–off of l_1 adaptive controller: a linear matrix inequality approach, *Proceedings of IAA Guidance, Navigation and Control Conference and Exhibit*, Publisher, Honolulu (USA), pp. 1–12.

Li, Y., Qiang, S., Zhuang, X. & Kaynak, O. (2004). Robust and adaptive backstepping control for nonlinear systems using rbf neural networks, *IEEE Transactions on Neural Networks* Vol. 15(No. 3): 693–701.

Liu, C. & Tong, S. (2010). Fuzzy adaptive decentralized control for nonlinear systems with unknown high-frequency gain sign based on k–filter, *Proceedings of Control and Decision Conference*, Publisher, China, pp. 1576–1581.

Nakanishi, J., Farrell, J. & Schaal, S. (2005). Composite adaptive control with locally weighted statistical learning, *Neural Networks* Vol. 18(No. 1): 71–90.

Park, B., Yoo, S., Park, J. & Choi, Y. (2009). Adaptive neural sliding mode control of nonholonomic wheeled mobile robots with model uncertainty, *IEEE Transactions on Control Systems Technology* Vol. 17(No. 1): 201–214.

Psillakis, H. (2010). Further results on the use of nussbaum gains in adaptive neural network control, *IEEE Transactions on Automatic Control* Vol. 55(No. 12): 2841–2846.

Ren, B., Ge, S., Lee, T. & Su, C. (2008). Adaptive neural control for uncertain nonlinear systems in pure-feedback form with hysteresis input, *Proceedings of the 47th IEEE Conference on Decision and Control*, Publisher, Cancun (Mexico), pp. 86–91.

Royden, H. (1988). *Real Analysis*, Prentice Hall Upper Saddle River, New Jersey.

Slotine, J. & Li, W. (1991). *Applied Nonlinear Control*, Prentice Hall Englewood Cliffs, New Jersey.

Su, C., Feng, Y., Hong, H. & Chen, X. (2009). Adaptive control of system involving complex hysteretic nonlinearities: a generalised prandtl-ishlinskii modelling approach, *International Journal of Control* Vol. 82(No. 10): 1786–1793.

Tong, S., Liu, C. & Li, Y. (2010). Fuzzy–adaptive decentralized output–feedback control for large–scale nonlinear systems with dynnamical uncertainties, *IEEE Transaction on Fuzzy Systems* Vol. 18(No. 5): 1–1.

Tong, S., Tang, J. & Wang., T. (2000). Fuzzy adaptive control of multivariable nonlinear systems, *Fuzzy Sets and Systems* Vol. 111(No. 2): 153–167.

Wang, X., Su, C. & Hong, H. (2004). Robust adaptive control of a class of nonlinear systems with unknown dead-zone, *Automatica* Vol. 40(No. 3): 407–413.

Wen, C., Zhou, J. & Wang, W. (2009). Decentralized adaptive backst epping stabilization of interconnected systems with dynamic input and output interactions, *Automatica* Vol. 45(No. 1): 55–67.

Yao, B. (1997). High performance adaptive robust control of nonlinear systems: a general framework and new schemes, *Proceedings of the 36th Conference on Decision and Control*, Publisher, San Diego (California, USA), pp. 2489–2494.

Yao, B. & Tomizuka, M. (1994). Robust adaptive sliding mode control of manipulators with guaranteed transient performance, *Proceedings of the American Control Conference*, Publisher, Baltimore (Maryland, USA), pp. 1176–1180.

Yao, B. & Tomizuka, M. (1997). Adaptive robust control of siso nonlinear systems in a semi-strict feedback form, *Automatica* Vol. 33(No. 5): 893–900.

Yousef, H. & Wahba, M. (2009). Adaptive fuzzy mimo control of induction motors, *Expert Systems with Applications* Vol. 36(No. 3): 4171–4175.

Zhang, T. & Ge, S. (2009). Adaptive neural network tracking control of mimo nonlinear systems with unknown dead zones and control directions, *IEEE Transactions on Neural Networks* Vol. 20(No. 3): 483–497.

Zhou, J., Wen, C. & Wang, W. (2009). Adaptive backstepping control of uncertain systems with unknown input time–delay, *Automatica* Vol. 45(No. 6): 1415–1422.

Zhou, J., Wen, C. & Zhang, Y. (2004). Adaptive backstepping control of a class of uncertain nonlinear systems with unknown backlash-like hysteresis, *IEEE Transactions on Automatic Control* Vol. 49(No. 10): 1751–1757.

6

A Robust Motion Tracking Control of Piezo-Positioning Mechanism with Hysteresis Estimation

Amir Farrokh Payam,
Mohammad Javad Yazdanpanah and Morteza Fathipour
Department of Electrical and Computer Engineering, University of Tehran, Tehran, Iran

1. Introduction

Piezoelectric actuators are the most suited actuation devices for high precision motion operations in the positioning tasks include miro/nano-positioning [1]. These actuators have unlimited motion resolution and posses some advantages such as ignorable friction, noiseless, zero backlash and easy maintenance, in comparison with the conventional actuated systems which are based on the sliding or revolute lower pairs [2].

Producing large forces, fast response and high efficiency are major advantages of piezoelectric actuators. But, it has some drawbacks such as hysteresis behavior, drift in time, temperature dependence and vibration effects. Molecular friction at sites of materials imperfections due to domain walls motion is the general cause of hysteresis in piezoelectric materials [3]. The hysteresis is a major nonlinearity for piezo-actuators and often limits system performance via undesirable oscillations or instability. Therefore, it is difficult to obtain an accurate trajectory tracking control. Numerous mathematical methods have been proposed to analyze the hysteresis behavior of piezoelectric actuators. These studies may be categorized in asymmetrical and symmetrical methods. The asymmetrical types of hysteretic models include polynomial model [4], Preisach's model [5], neural network model [6] and Karasnoselskii and Pokrovskii [7]. The symmetrical types of hysteretic model include Duhem model [8], Bouc-Wen model [9] and Lugre model [9].

The asymmetrical methods establish the nonlinear relations between the input and output based on the measured input/output data sets. Because superposition of a basic hysteresis operator is a fundamental principle in these models, they are also called to operator based model. Although, an operator based model may give a good match with experimental data, the dynamics of the piezoelectric material is not formulated in these modeling methods and model parameter identification and implementation is more difficult in this case. The symmetrical methods employ nonlinear differential equations in order to describe hysteresis. In this case, the dynamics of the piezoelectric materials are described but the non-symmetric hysteresis is not modeled. However these models are more tractable for control design. In order to include the hysteresis effect and compensating its effect, Lugre model [10] is analyzed and studied in this chapter.

There are several control methods to overcome the above mentioned errors and increase the tracking control precision of the piezoelectric actuators. Some of these methods are PI and PID controller, fuzzy controller [11], adaptive RFNN [12], feed-forward model reference control method [13], adaptive hysteresis inverse cascade with the plant [14], reinforcement discrete neuro-adaptive controller [15], adaptive wavelet neural network controller [16], nonlinear observer-based sliding-mode controller [17] an adaptive backstepping controller [10, 18], robust motion tracking controller based on sliding-mode theory [19] and continuous time controller based on SMC and disturbance observer [20]. In some of these works, a complex inverse hysteresis model has been adopted to overcome the nonlinear hysteresis effect. Also, in the methods based on the neural network approach, to ensure the error is bounded, it is assumed that the system states must be inside a compact set. Moreover, robustness against parameters uncertainties and external disturbances is the other problem encounter the control methods presented for piezoelectric actuator.

In this chapter, a robust motion tracking control strategy in combination with the hysteresis force observer is designed and investigated for the piezoelectric actuator. The presented controller is robust against the unknown or uncertain system parameters and can estimate the hysteresis force with its estimation property. This control strategy is established based on the lumped parameter dynamic model. Using Lyapanouv stability analysis, the stability analysis of the overall control and observer system is performed. Furthermore, the validity and effectiveness of the designed methodology is investigated by numerical analysis and its results obtained are compared with those of [19].

Section 2 contains the model explanation of piezo-positioning mechanism. A hysteresis model for piezoelectric systems based on Lugre model is discussed in section 3. Design procedure of the developed controller is presented in section 4. Simulation analysis and results obtained are presented in section 5. Finally, section 6 includes the conclusion of this chapter.

2. Model of piezo-positioning mechanism with hysteresis

The lumped parameter dynamic model of the piezo-driven mechanism is written as [10, 16]:

$$M\ddot{x} + D\dot{x} + F_H + F_L = u \tag{1}$$

Where M is the mass of the controlled piezo-positioning mechanism, D is the linear friction coefficient of the piezo-driven system, F_L denotes the external load, F_H is the hysteresis friction force function, $x(t), \dot{x}(t)$, and $\ddot{x}(t)$ denote the piezoelectric displacement, velocity and acceleration, respectively and u is the applied voltage to the piezo-positioning mechanism.

A block diagram of the model (1) is depicted in Fig.1. Note that the simulation program used in this chapter is C++.

Noted that:

$$|F_H(t)| \le \delta F_{H0}, |\dot{F}_H(t)| \le \delta F_{H1}, |\ddot{F}_H(t)| \le \delta F_{H2},, |F_L(t)| \le \delta F_{L0}\ |\dot{F}_L(t)| \le \delta F_{L1}, |\ddot{F}_L(t)| \le \delta F_{L2} \tag{2}$$

Where $\delta F_{Hi} \in \Re^1, i = 0,1,2$ and $\delta F_{Li} \in \Re^1, i = 0,1,2$ denote the known upper bounds.

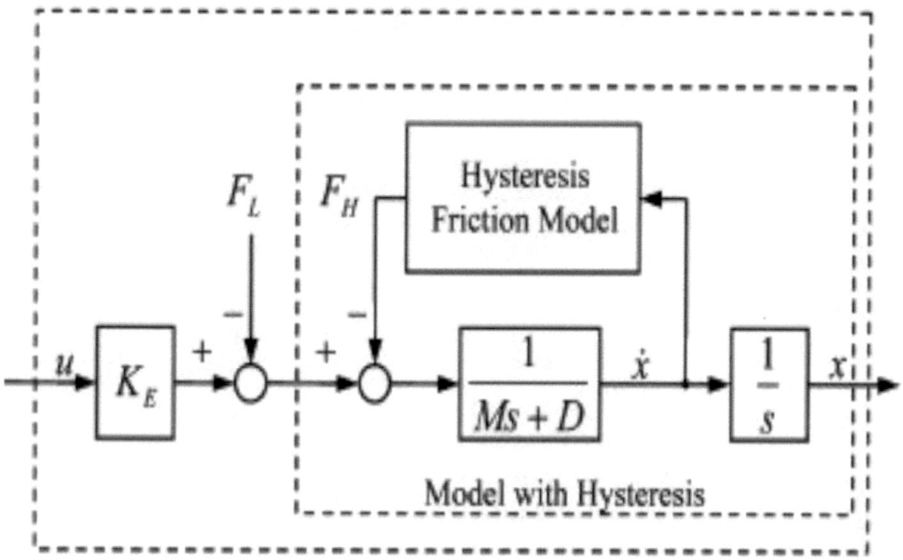

Fig. 1. Model of piezo-positioning mechanism with hysteresis [10].

3. Lugre hysteresis model for piezo-positioning system

This chapter uses Lugre model to describe nonlinear hysteretic curve of piezoelectric systems. The mathematical equation is as follows [10,16]:

$$F_H = \sigma_0 z - \sigma_1 \frac{1}{g(\dot{x})} z|\dot{x}| + (\sigma_1 + \sigma_2)\dot{x} \tag{3}$$

$$\dot{z} = \dot{x} - \frac{|\dot{x}|}{g(\dot{x})} z \tag{4}$$

Where z interpreted as the contact force applied voltage average bristle coefficient, $\sigma_0, \sigma_1, \sigma_2$ are positive constants and generally can be equivalently interpreted a bristle stiffness, damping and viscous-damping coefficients, respectively. Moreover, the function $g(\dot{x})$ denotes the Stribeck effect curve described by:

$$\sigma_0 g(\dot{x}) = f_C + (f_S - f_C) e^{-(\dot{x}/\dot{x}_S)^2} \tag{5}$$

Where f_C is the coloumb friction level, f_S is the level of stiction force and $\dot{x}_S$ is the Stribeck velocity. The function $g(\dot{x})$ is positive and constant [16]. As depicted in [10], the following lemma is held.

Lemma 1. Consider nonlinear dynamics system of (4). For any piecewise continuous signal x and $\dot{x}$, the output $z(t)$ is bounded.

Substituting the (3) in (1) and arranging the expression, the following dynamics equation is obtained for the piezo-positioning mechanism with Lugre hysteresis model:

$$\ddot{x} = \frac{1}{M}u - \frac{1}{M}\left[(\sigma_0 z + F_L) - \sigma_1 \frac{z|\dot{x}|}{g(\dot{x})} + (\sigma_1 + \sigma_2)\dot{x}\right] - \frac{D}{M}\dot{x} \tag{6}$$

4. Design controller

In this section we design a robust control strategy for the system of (1), to asymptotically estimate the hysteresis parameter of equation (3). To facilitate the design process, we assumed that the hysteresis force is dependent only on the time and F_H and its first two time derivatives remain bounded for all times. Using (6) we can rewrite equation (1) as:

$$\ddot{x} = \frac{u}{M} - \frac{F}{M} - \frac{D}{M}\dot{x} \tag{7}$$

Where $F = (\sigma_0 z + F_L) - \sigma_1 \frac{z|\dot{x}|}{g(\dot{x})} + (\sigma_1 + \sigma_2)\dot{x}$ is a combination of an unknown friction hysteresis force function and external load, which must be estimated.

The dynamics of a piezo-positioning system can be represented by the following equation:

$$\bar{M}\ddot{x} + \bar{D}\dot{x} + F - u + (\Delta M\ddot{x} + \Delta D\dot{x}) = 0 \tag{8}$$

Where $\bar{M}$ and $\bar{D}$ are the nominal parameter values of the mass and linear friction coefficient and ΔM and ΔD are the parametric errors between real value and nominal value of the uncertain parameters of the system and modeled as:

$$|\Delta M| \le \delta M \tag{9}$$

$$|\Delta D| \le \delta D \tag{10}$$

Where δM and δD are the bounds of system parameters. The position tracking error signal is defined as:

$$e = x_d - x \tag{11}$$

Where $x_d \in \Re^1$ denotes the desired position trajectory. The desired piezo-position trajectory and its first three time derivatives are assumed to be constrained by the following:

$$|x_d| < \zeta_{d0}, |\dot{x}_d| < \zeta_{d1}, |\ddot{x}_d| < \zeta_{d2}, |\dddot{x}_d| < \zeta_{d3} \tag{12}$$

Where the ζ_{di}'s denote known positive constants. Also, the filtered tracking error signal $r(t) \in \Re^1$ is defined as:

$$r = \dot{e} + \alpha e \tag{13}$$

Where $\alpha \in \Re^1$ is a positive constant control gain.

The system dynamics of (8) are rewritten in terms of the filtered tracking error signal $r(t)$ as follows:

$$\dot{r} = (\ddot{x}_d + \alpha\dot{e}) + \frac{\bar{D}}{\bar{M}}\dot{x} + \frac{F}{\bar{M}} + \frac{1}{\bar{M}}(\Delta M\ddot{x} + \Delta D\dot{x}) - \frac{u}{\bar{M}} \tag{14}$$

We define χ as:

$$\chi = \frac{1}{\bar{M}}(\Delta M\ddot{x} + \Delta D\dot{x}) \leq \frac{1}{\bar{M}}(\delta M|\ddot{x}| + \delta D|\dot{x}|) = \frac{1}{\bar{M}}\beta_1(x) \tag{15}$$

Another filtered tracking error signal $s(t) \in \Re^1$ is defined as:

$$s = \dot{r} + \beta r \tag{16}$$

Where β is a constant positive parameter.

Based on the developed error system and ensuring the stability analysis, the following control input signal for the system (14) is designed:

$$u = \bar{M}(\ddot{x}_d + \alpha\dot{e}) + \bar{D}\dot{x} + \hat{F} + \beta_1(x)\text{sgn}(s) \tag{17}$$

Where $\hat{F} \in \Re^1$ represents the estimation of F and is obtained by the following observer:

$$\dot{\hat{F}} = -(k_1 + \beta)\hat{F} + k_1\beta r + \rho\,\text{sgn}(r) \tag{18}$$

Where sgn(.) is the standard signum function, and $k_1, \rho \in \Re^1$ are positive constants.

Noted that the equation (18) for $\hat{F}$ is a stable linear system with the disturbance term $k_1\beta r + \rho\,\text{sgn}(r)$.

To facilitate the dynamic system stability analysis, the auxiliary disturbance signal $\eta(t) \in \Re^1$ is defined by:

$$\eta = \dot{F} + (k_1 + \beta)F \tag{19}$$

Due to the boundness of F and its first two time derivatives, it is understand that $\eta(t), \dot{\eta}(t) \in L_\infty$.

Based on the stability analysis, the constant ρ should be chosen to satisfy the following inequality:

$$\rho \geq |\eta(t)| + \frac{1}{\beta}|\dot{\eta}(t)| \tag{20}$$

By substituting (17) in (14), and simplify the obtained expression, the dynamics of $\dot{r}$ is achieved as:

$$\dot{r} = F - \hat{F} + \frac{1}{\bar{M}}(\chi_1(x) - \beta_1(x)\text{sgn}(s)) \tag{21}$$

Where $\chi_1(x) = \Delta M\ddot{x} + \Delta D\dot{x}$.

Now, using (21), the time derivative of (16) is obtained as:

$$\dot{s} = \dot{F} - \dot{\hat{F}} + \frac{1}{\bar{M}}(\dot{\chi}_1(x) - \dot{\beta}_1(x)\text{sgn}(s)) + \beta(F - \hat{F} + \frac{1}{\bar{M}}(\chi_1(x) - \beta_1(x)\text{sgn}(s))) \tag{22}$$

Substituting (18) and (19) in (22), gives:

$$\dot{s} = \eta - k_1 s - \rho\,\text{sgn}(r) + \frac{1}{\bar{M}}(\dot{\chi}_1(x) - \dot{\beta}_1(x)\text{sgn}(s)) + \frac{\beta + k_1}{\bar{M}}(\chi_1(x) - \beta_1(x)\text{sgn}(s)) \tag{23}$$

Remark 1: If $s(t) \in L_\infty$, then $r(t), \dot{r}(t) \in L_\infty$ and if $s(t)$ is asymptotically regulated, then $r(t), \dot{r}(t)$ are also, asymptotically regulated.

5. Stability analysis

Theorem 1. For the dynamics of (7), the designed controller of (17) and (18) guarantees the global asymptotetic piezo-driven position tracking in the sense that:

$$\lim_{t\to\infty} e(t) = 0 \tag{24}$$

And global asymptotetic estimation of hysteresis in the sense that:

$$\lim_{t\to\infty} [\hat{F} - F] = 0 \tag{25}$$

With the constant ρ satisfies the condition of (20).

Proof

A non-negative, scalar function $V(t) \in \Re^1$ is defined as:

$$V = \frac{1}{2}s^2 \tag{26}$$

Tacking time derivative of (26) gives:

$$\dot{V} = -k_1 s^2 + (\dot{r} + \beta r)(\eta - \rho\,\text{sgn}(r)) + \frac{s}{\bar{M}}(\dot{\chi}_1(x) - \dot{\beta}_1(x)\text{sgn}(s)) + \frac{\beta + k_1}{\bar{M}} s(\chi_1(x) - \beta_1(x)\text{sgn}(s)) \tag{27}$$

From (15) it is clear that:

$$\frac{1}{\bar{M}}(\dot{\chi}_1(x) - \dot{\beta}_1(x)\text{sgn}(s)) + \frac{\beta + k_1}{\bar{M}}(\chi_1(x) - \beta_1(x)\text{sgn}(s)) \le 0 \tag{28}$$

Where $\dot{\beta}_1(x) = \delta M|\dddot{x}| + \delta D|\ddot{x}|$.

Hence:

$$\dot{V} \le -k_1 s^2 + (\dot{r} + \beta r)(\eta - \rho\,\text{sgn}(r)) \tag{29}$$

After integrating both sides of (29), the following inequality is obtained:

$$V(t)-V(t_0) \le -k_1\int_{t_0}^{t} s^2(\sigma)d\sigma + \left[|r(t)||\eta(t)| - \rho|r(t)|\right] + \left[\int_{t_0}^{t} \beta|r(\sigma)|\left(|\eta| + \frac{1}{\beta}\left|\frac{d\eta(\sigma)}{d\sigma}\right| - \rho\right)d\sigma\right] + \zeta_0 \quad (30)$$

Where $\zeta_0 \in \Re^1$ is a positive constant, defined by:

$$\zeta_0 = |r(t_0)||\eta(t_0)| + \rho|r(t_0)| \quad (31)$$

After applying the (20) to the bracketed term of (30), $V(t)$ can be upper bounded as follows:

$$V(t) \le V(t_0) - k_1\int_{t_0}^{t} s^2(\sigma)d\sigma + \zeta_0 \quad (32)$$

It is deduced from (32) and (26) that $V(t) \in L_\infty$ and $s(t) \in L_\infty$, respectively. Finally utilizing remark 1, $r(t), \dot{r}(t), e(t), \dot{e}(t), \dot{s}(t) \in L_\infty$.

Equation (32) rearrange as:

$$k_1\int_{t_0}^{t} s^2(\sigma)d\sigma \le V(t_0) - V(t) + \zeta_0 \quad (33)$$

From the fact that $V(t)$ is non-negative and equation (33), it can be deduced that $s(t) \in L_2$. Because of $s(t) \in L_\infty \cap L_2$ and $\dot{s}(t) \in L_\infty$, we can use the Barbalat's Lemma [21] to summarized that:

$$\lim_{t\to\infty} s(t) = 0 \quad (34)$$

Therefore:

$$\lim_{t\to\infty} r(t), \dot{r}(t), e(t), \dot{e}(t) = 0 \quad (35)$$

6. Simulation results

In this section, simulation results are presented to investigate the performance of the presented method for piezoelectric actuator. First, we test the controller response with the nominal value of piezoelectric actuator, when $F_L = 0.3\sin(t)$ and initial values are: $x(0) = 0, \dot{x}(0) = 0, z(0) = 0$. The objective of the positioning is to drive the displacement signal x to track the reference trajectory which is shown in Fig.2.

The parameters of piezo-positioning mechanism are given in Table 1.

$\sigma_0 = 50000N/m$	$\sigma_1 = \sqrt{5\times 10^4}Ns/m$	$\sigma_2 = 0.4Ns/m$
$f_C = 1N$	$f_S = 1.5N$	$\dot{x}_S = 0.001m/s$
$M = 1kg$	$D = 0.0015Ns/m$	$F_L = 0.3\sin(t)$

Table 1. Piezoelectric Parameters

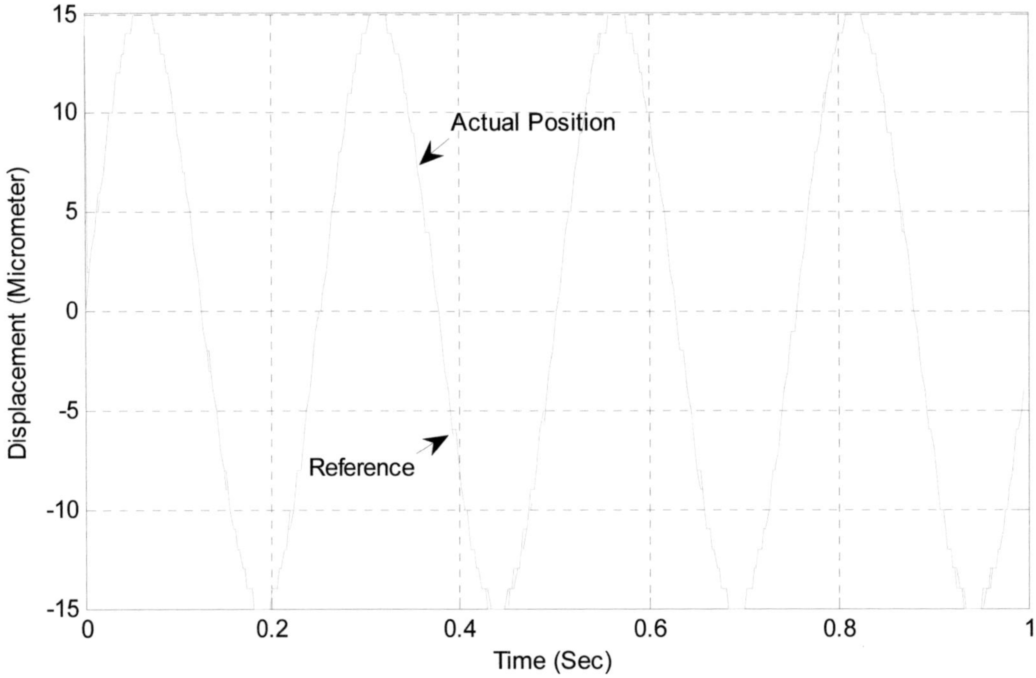

Fig. 2. Desired and Actual Piezoelectric displacement.

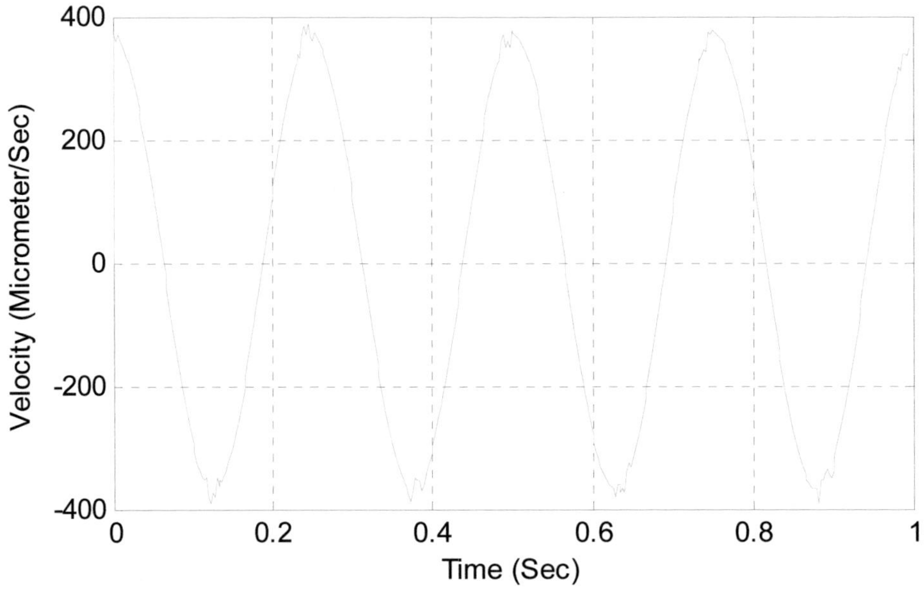

Fig. 3. Piezoelectric velocity.

As depicted from the results, especially Fig.4, the positioning mechanism has a good accuracy and the error between displacement and reference signal at least 4 orders is smaller than the actual signal which means that the error is about %0.01.

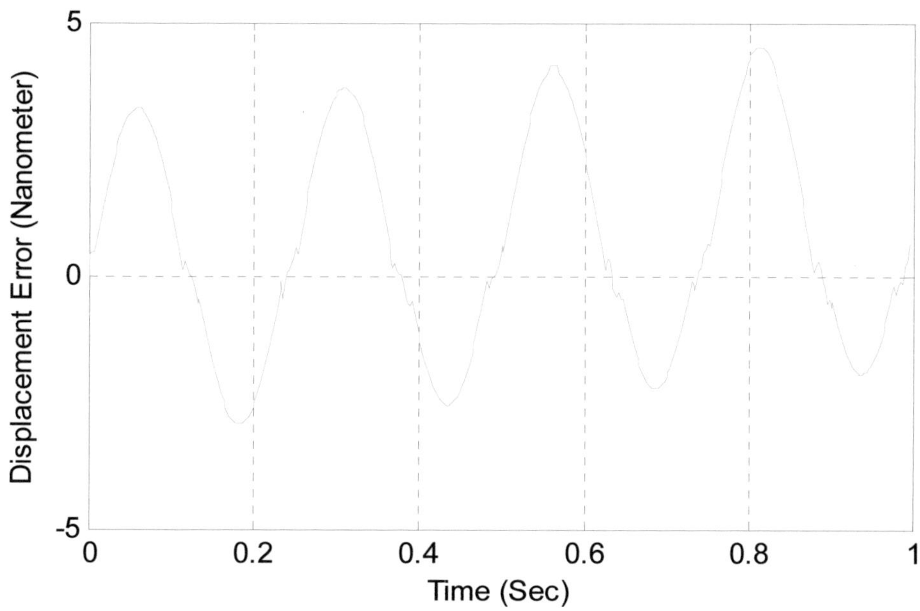

Fig. 4. Error between actual & desired displacement.

For the purpose of study the behavior of the controller in the presence of parameter uncertainties, by considering the $\Delta M = 0.1\bar{M}$, $\Delta D = 0.1\bar{D}$ and $\Delta F_L = 0.4$ and using $x(0) = 15(\mu m), \dot{x}(0) = 0, z(0) = 0$ as the initial values, we have performed another test. The result of this test is compared with the result of the method presented in [19]. In the simulation we consider the control gain α is 2000.

As it can be seen from Fig.5 and Fig.7, the presented method has an acceptable response and the positioning error is in about %0.25. Also, the hysteresis and disturbance voltage and its estimation are shown in Fig.9. As it can be seen from this result, the hysteresis identifier can estimate the hysteresis voltage precisely. For the purpose of comparison the accuracy of the proposed method with the recently proposed method in [19], we simulate the piezo-positioning mechanism with the method of [19]. The displacement error of [19] is shown in Fig.10. Noted that in the simulation of [19] we use these control gains: $k_p = 10^5, k_v = 9000, k_s = 50$ and $\alpha = 10$. Comparison of Fig.7 and Fig.10 depicts that the error in the presented method is smaller than the method of [19]. Also, it is clear that in addition of smaller number of control gains, the gain of the controller in the presented method in comparison with [19] is very smaller. Experimental implementation of the high gain needs more complexity and also it may be generate noise. Although method of [19] is

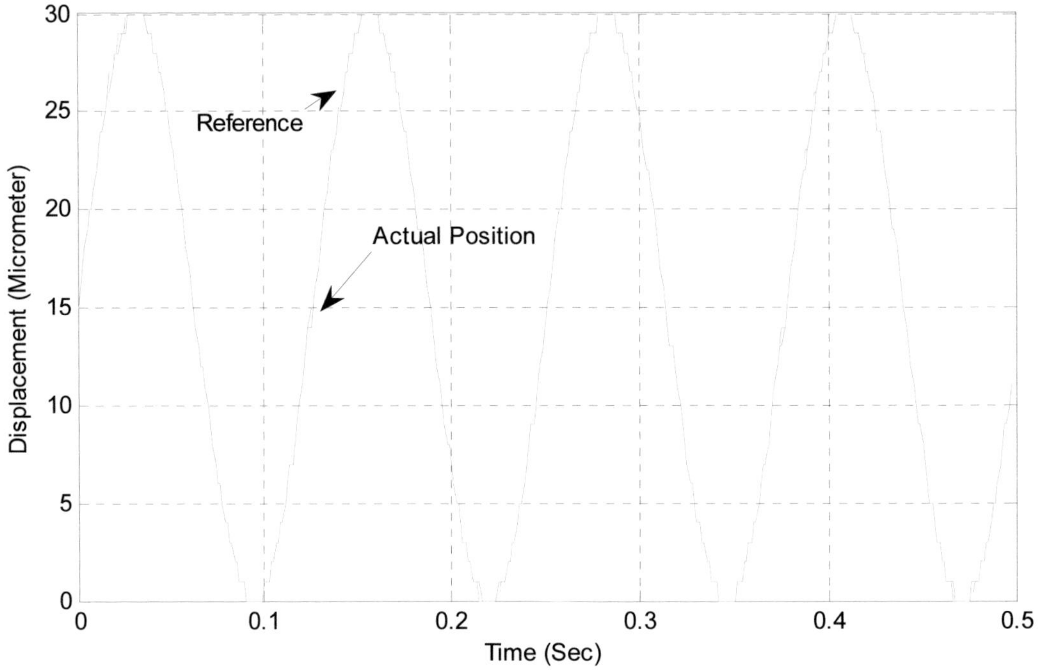

Fig. 5. Displacement of piezoelectric.

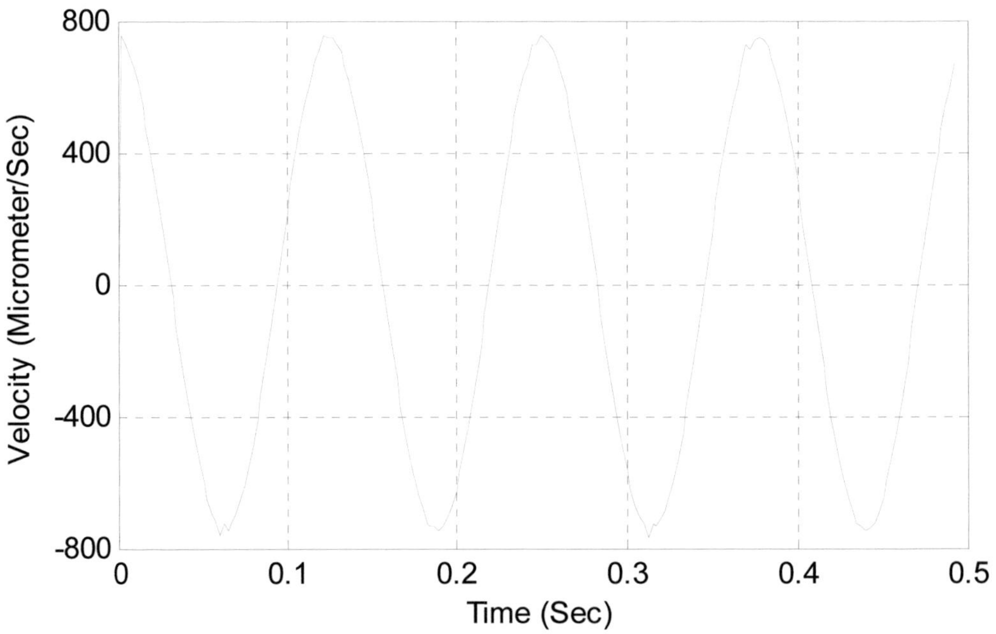

Fig. 6. Velocity of piezoelectric.

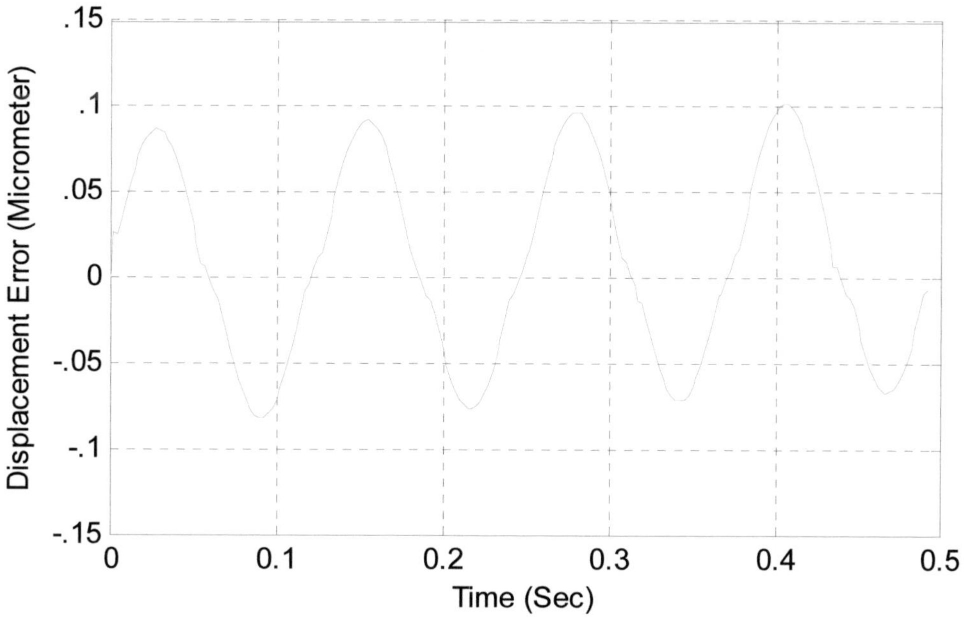

Fig. 7. Error between actual & desired displacement.

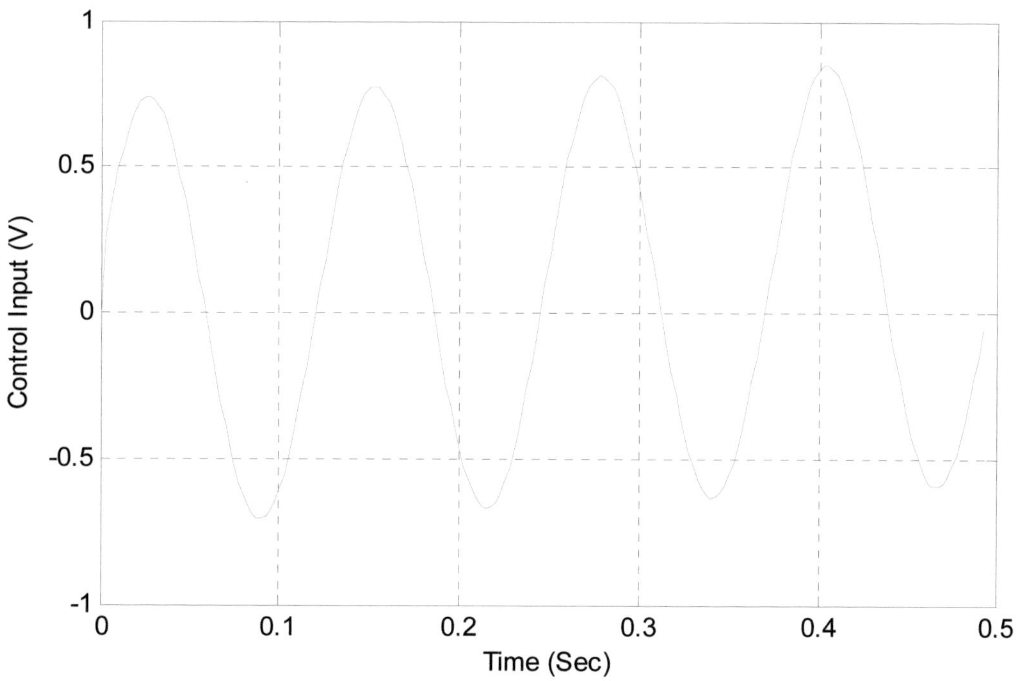

Fig. 8. Control input.

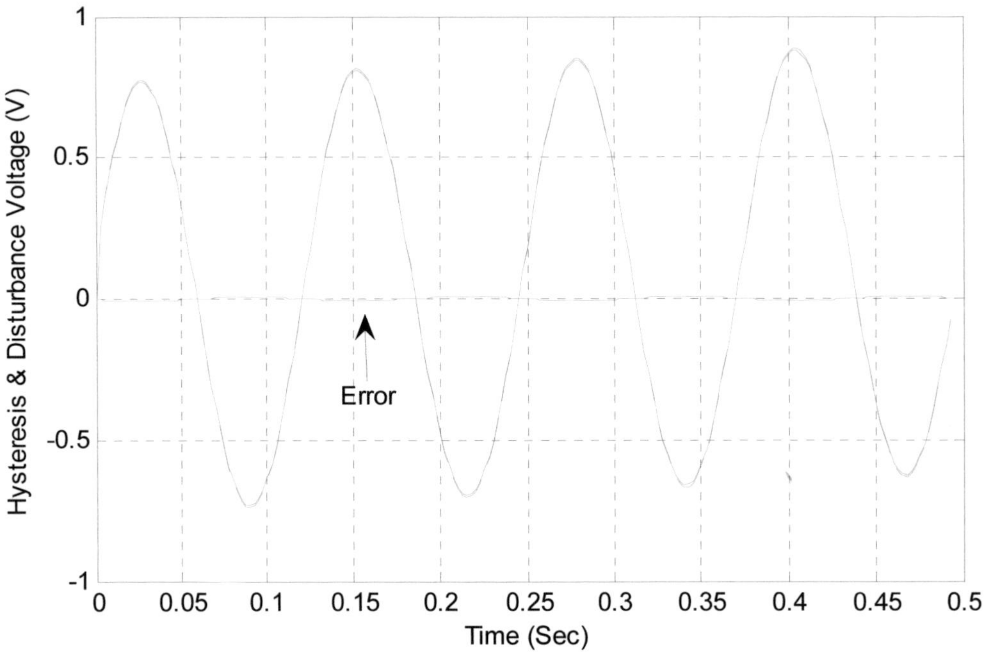

Fig. 9. Estimated & actual values of hysteresis and disturbance voltages and error between them.

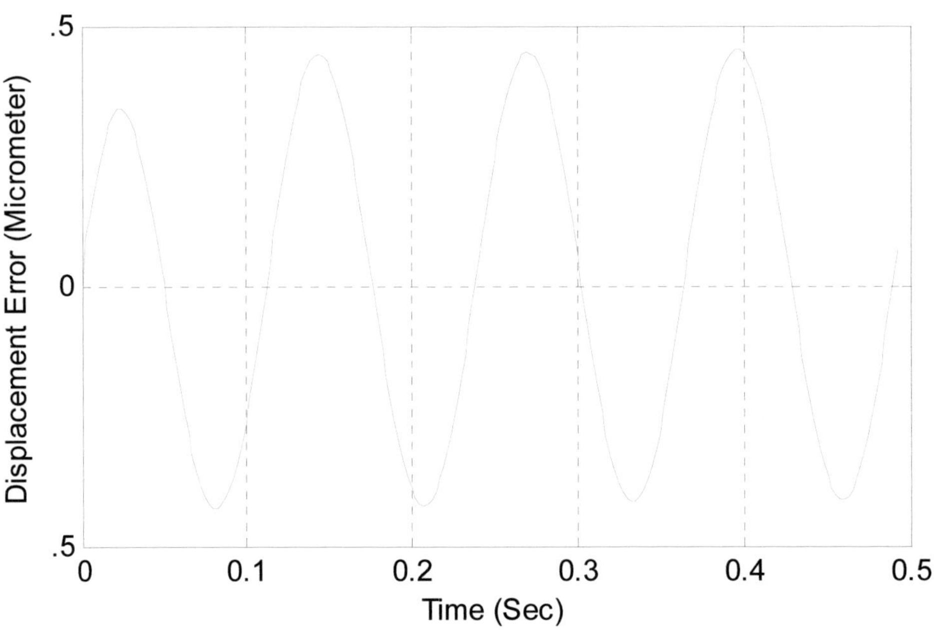

Fig. 10. Displacement Error of the method presented in [19].

robust against disturbances and uncertainties and in comparison with other methods has higher precision, it needs high value gains to perform this task. While the proposed method can perform these tasks with lower cost and also it has the capability of hysteresis estimation.

For the comparison between the presented method and the method of [19] another simulation test is carried out. In this case $\Delta M = 0.2\bar{M}$, $\Delta D = 0.2\bar{D}$ and $\Delta F_L = 0.4$. The result is shown in Fig.11.

As it can be depicted the presented method has the better response and using the proposed controller the error is decreased. Note that in this case the reference is similar to the Fig.5.

The last simulation is devoted to the reference signal of Fig.2 by considering $\Delta M = 0.2\bar{M}$, $\Delta D = 0.2\bar{D}$ and $\Delta F_L = 0.4$. Also in this case the error of the proposed controller is less than the method of [19].

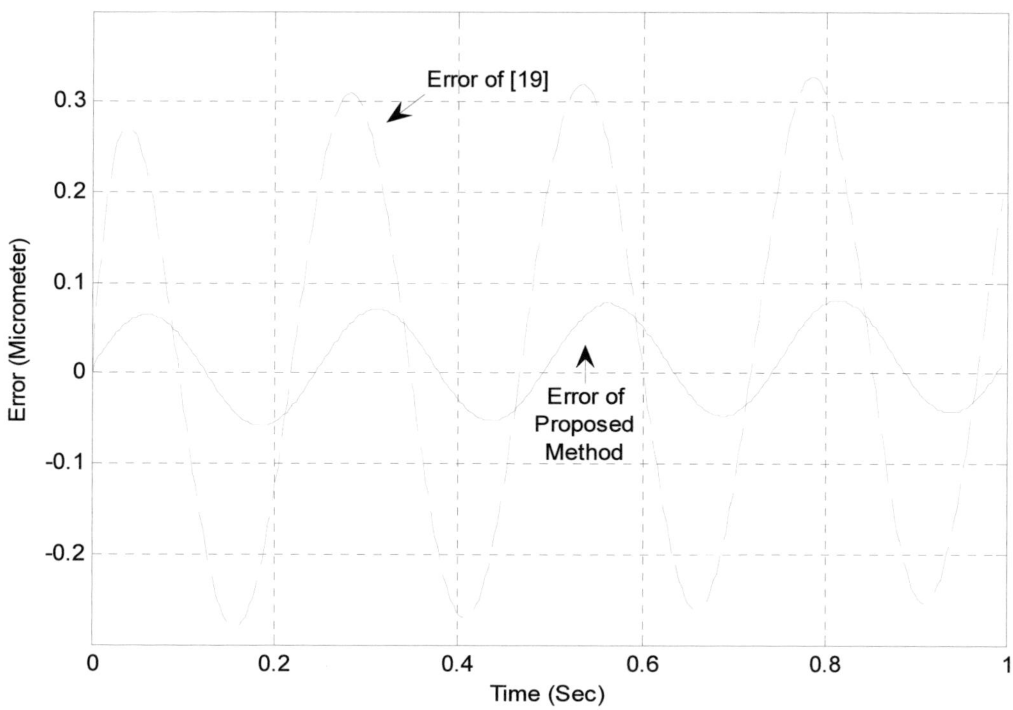

Fig. 11. Comparison between the displacement error of the presented controller and the controller proposed in [19].

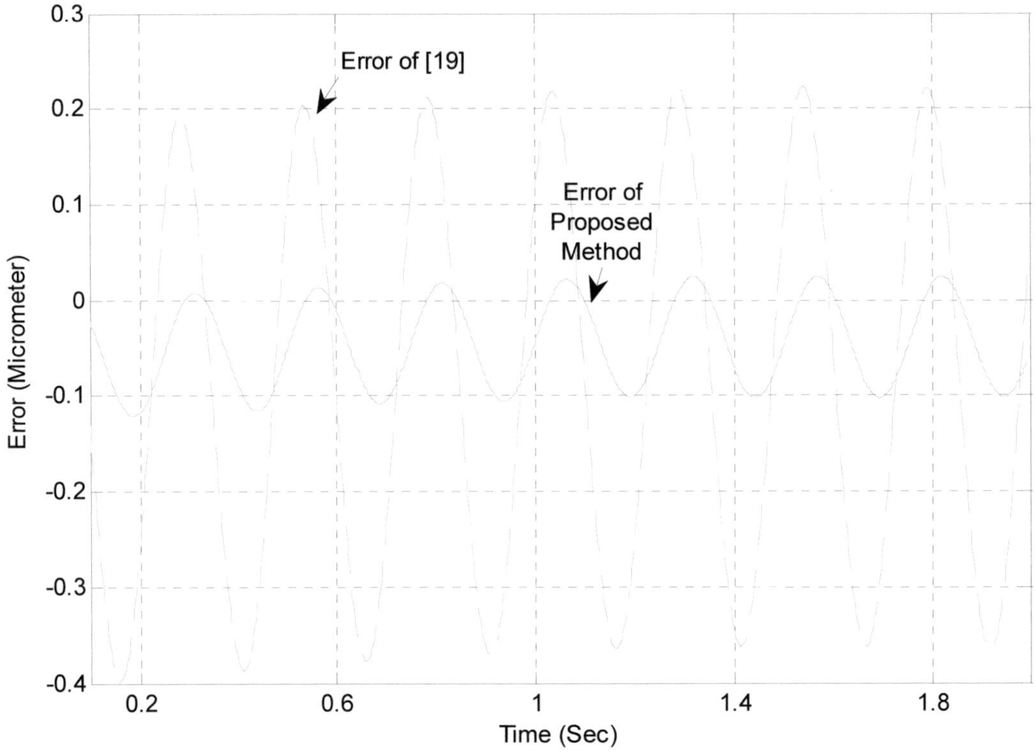

Fig. 12. Comparison between the displacement error of the presented controller and the controller proposed in [19].

7. Conclusion

A robust tracking motion controller is designed to control a positioning of piezoelectric actuator system with hysteresis phenomenon. The Lugre hysteresis model is used to model the nonlinearities in the system under study. Using the presented controller, we can estimate the nonlinear hysteresis and disturbances imposed to the piezoelectric actuator and compensate their effects. The performance and efficiency of the designed controller for a positioning system is compared with recently proposed robust method. The results obtained depict the validity and performance of the presented approach.

8. References

[1] Spanner, K. and S. Vorndran, 2003. Advances in piezo-nanopositioning technology. Proc. IEEE/ASME int. conf. advanced intelligent mechatronics, Kobe, Japan.

[2] Yi B.J., G.B. Chung, H.Y. Na, W.K. Kim and I.H. Suh, 2003. Design and experiment of a 3-DOF parallel micromechanism utilizing flexurehinges. IEEE Trans Robotics Automation, 19(4): 604–12.

[3] EI Rifai, O.M., 2002. Modeling and control of undesirable dynamics in atomic force microscopes. PhD dissertation, MIT, February.

[4] Hung, X.C. and D.Y. Lin, 2003. Tracking control of a piezoelectric actuator based on experiment hysteretic modeling. Proc. Of the IEEE/ASME Int. Conf. Advanced Intelligent Mechatronics, Kobe, Japan.

[5] Ge, P. and M. Jouaneh, 1997. Generalized priesach model for hysteresis nonlinearity of piezoceramic actuators. Precision Eng., 20: 99-111.

[6] Richter, H., E.A. Misawa, D.A. Lucca and H. Lu, 2001. Modeling nonlinear behavior in a piezoelectric actuator. Prec. Eng. J., 25: 128-137.

[7] Krasnosellskii, M.A. and A.V. Pokrovskii, 1989. Systems with hysteresis, Springer-Verlag, Berlin.

[8] Stepanenko, Y. and C.Y. Su, 1998. Intelligent control of piezoelectric actuators. Proc. IEEE Conf. Decision & Control, 4234-4239.

[9] Wen, Y.K. 1976. Method for random vibration of hysteresis system. J. Eng. Mechanics Division, 102: 249-263.

[10] Zhou, J., C. Wen and C. Zhang, 2007. Adaptive backstepping control of piezo-positioning mechanism with hysteresis. Trans. CSME, 31(1): 97-110.

[11] Yavari, F., M.J. Mahjoob, and C. Lucas, 2007. Fuzzy control of piezoelectric actuators with hysteresis for nanopositioning. 13th IEEE/IFAC Int. Conf. on Methods and Models in Automation and Robotics, 685-690, August.

[12] Xu, J.H. 1993. Neural network control of a piezo tool positioner. Canadian Conf. Elect. & Comp. Eng., 1: 333-336.

[13] Jung, S.B. and S.W. Kim, 1994. Improvement of scanning accuracy of pzt piezoelectric actuator by feed-forward model-reference control. Precision Engineering, 16: 49-55.

[14] Tao, G. and P.V. Koktokovic, 1995. Adaptive control of plants with unknown hysteresis. IEEE Transaction on Auotomatic Control, 40: 200-212.

[15] Huwang, L. and C. Jans, 2003. A reinforcement discrete neuro-adaptive control of unknown piezoelectric actuator systems with dominant hysteresis. IEEE Trans. Neural Networks, 14: 66-78.

[16] Lin, F.J., H.J. Shieh, and P.K. Huang, 2006. Adaptive wavelet neural network control with hystereis estimation for piezo-positioning mechanism. IEEE Trans. Neural Networks, 17: 432-444.

[17] Jan, C. and C.L. Hwang, 2004. A nonlinear observer-based sliding-mode control for piezoelectric actuator systems: theory and experiments. J Chinese Inst. Eng.,27(1): 9-22.

[18] Shieh H.J., F.J. Lin, P.K. Huang, and L.T. Teng, 2004. Adaptive tracking control solely using displacement feedback for a piezo-positioning mechanism. IEE Proc Control Theory Appl,151(5): 653–60.

[19] Liaw, H.C., B. Shirinzadeh, and J. Smith, 2008. Robust motion traking control of piezo-driven flexure-based four bar mehanism for micro/nano manipulation. Mechatronics, 18: 111-120.

[20] Yannier, S. and A. Sabanovic, 2007. Continuous Time Controller Based on SMC and Disturbance Observer for Piezoelectric Actuators. Int. Rev. Elect. Eng (I.R.E.E.), 3 (6).

[21] Spooner, J.T., M. Maggiore, R. Ordonez, and K.M. Passino, 2002. Stable Adaptive Control and Estimation for Nonlinear Systems. John Wiley and Sons, Inc, NY.

7

Nonlinear Observer-Based Control Allocation

Fang Liao[1], Jian Liang Wang[2] and Kai-Yew Lum[1]
[1]*National University of Singapore*
[2]*Nanyang Technological University*
Singapore

1. Introduction

Control allocation is the process of mapping virtual control inputs (such as torque and force) into actual actuator deflections in the design of control systems (Benosman et al., 2009; Bodson, 2002; Buffington et al., 1998; Liao et al., 2007; 2010). Essentially, it is considered as a constrained optimization problem as one usually wants to fully utilize all actuators in order to minimize power consumption, drag and other costs related to the use of control, subject to constraints such as actuator position and rate limits. In the design of control allocation, full state information is required. However, in practice, states may not be measurable. Hence, estimation of these unmeasurable states becomes inevitable.

The unmeasurable states are generally estimated based on available measurements and the knowledge of the physical system. For linear systems, the property of observability guarantees the existence of an observer. Luenberger or Kalman observers are known to give a systematic solution (Luenberger, 1964). In the case of nonlinear systems, observability in general depends on the input of the system. In other words, observability of a nonlinear system does not exclude the existence of inputs for which two distinct initial states generate identical measured outputs. Hence, in general, observer gains can be expected to depend on the applied input (Nijmeijer & Fossen, 1999). This makes the design of a nonlinear observer for a general nonlinear system a challenging problem. Although various results have been proposed over the past decades (Ahmed-Ali & Lamnabhi-Lagarrigue, 1999; Alamir, 1999; Besancon, 2007; Besancon & Ticlea, 2007; Bestle & Zeitz, 1983; Bornard & Hammouri, 1991; Gauthier & Kupka, 1994; Krener & Isidori, 1983; Krener & Respondek, 1985; Michalska & Mayne, 1995; Nijmeijer & Fossen, 1999; Teel & Praly, 1994; Tsinias, 1989; 1990; Zimmer, 1994), none of them can claim to provide a general solution with the same convergence properties as in the linear case.

Over the past decades, a variety of methods have been developed for constructing nonlinear observers for nonlinear systems (Ahmed-Ali & Lamnabhi-Lagarrigue, 1999; Alamir, 1999; Besancon, 2007; Besancon & Ticlea, 2007; Bestle & Zeitz, 1983; Bornard & Hammouri, 1991; Gauthier & Kupka, 1994; Krener & Isidori, 1983; Krener & Respondek, 1985; Michalska & Mayne, 1995; Nijmeijer & Fossen, 1999; Teel & Praly, 1994; Tsinias, 1989; 1990; Zimmer, 1994). They may be classified into optimization-based methods (Alamir, 1999; Michalska

& Mayne, 1995; Zimmer, 1994) and feedback-based methods (Bestle & Zeitz, 1983; Bornard & Hammouri, 1991; Gauthier & Kupka, 1994; Krener & Isidori, 1983; Krener & Respondek, 1985; Teel & Praly, 1994; Tsinias, 1989; 1990). Optimization-based methods obtain an estimate $\hat{x}(t)$ of the state $x(t)$ by searching for the best estimate $\hat{x}(0)$ of $x(0)$ (which can explain the evolution $y(\tau)$ over $[0, t]$) and integrating the deterministic nonlinear system from $\hat{x}(0)$ and under $u(\tau)$. These methods take advantage of their systematic formulation, but suffer from usual drawbacks of nonlinear optimization (like computation burden, local minima, and so on). Feedback-based methods can correct on-line the estimation $\hat{x}(t)$ from the error between the measurement output and the estimated output. These methods include linearization methods (Bestle & Zeitz, 1983; Krener & Isidori, 1983; Krener & Respondek, 1985), Lyapunov-based approaches (Tsinias, 1989; 1990), sliding mode observer approaches (Ahmed-Ali & Lamnabhi-Lagarrigue, 1999) and high gain observer approaches (Bornard & Hammouri, 1991; Gauthier & Kupka, 1994; Teel & Praly, 1994), and so on. Among them, linearization methods (Krener & Isidori, 1983) transform nonlinear systems into linear systems by change of state variables and output injection. It is applicable to a special class of nonlinear systems. Sliding mode observer approaches (Ahmed-Ali & Lamnabhi-Lagarrigue, 1999) is to force the estimation error to join a stabilizing variety. The difficulty is to find a variety attainable and having this property. High gain observer approaches (Besancon, 2007) use the uniform observability and weight a gain based on the linear part so as to make the linear dynamics of the observer error to dominate the nonlinear one. Due to the requirement of the uniform observability, these approaches can only be applied to a class of nonlinear systems with special structure. Interestingly, Lyapunov-based approaches (Tsinias, 1989; 1990) provide a general sufficient Lyapunov condition for the observer design of a general class of nonlinear systems and the proposed observer is a direct extension of Luenberger observer in linear case.

In this chapter, we extend the control allocation approach developed in (Benosman et al., 2009; Liao et al., 2007; 2010) from state feedback to output feedback and adopt the Lyapunov-type observer for a general class of nonlinear systems in (Tsinias, 1989; 1990) to estimate the unmeasured states. Sufficient Lyapunov-like conditions in the form of the dynamic update law are proposed for the control allocation design via output feedback. The proposed approach ensures that the estimation error and its rate converge exponentially to zero as $t \to +\infty$ and the closed-loop system exponentially converges to the stable reference model as $t \to +\infty$. The advantage of the proposed approach is that it is applicable to a wide class of nonlinear systems with unmeasurable states, and it is computational efficiency as it is not necessary to optimize the control allocation problem exactly at each time instant.

This chapter is organized as follows. In Section 2, the observer-based control allocation problem is formulated where the control allocation design is based on the estimated states which exponentially converge to the true states as $t \to +\infty$. In Section 3, the main result of the observer-based control allocation design is presented in the form of dynamic update law. An illustrative example is given in Section 4, followed by some conclusions in Section 5.

Throughout this chapter, given a real map $f(\mathbf{v}, \mathbf{w})$, $(\mathbf{v}, \mathbf{w}) \in \mathbb{R}^n \times \mathbb{R}^m$, $D_{\mathbf{v}} f(\mathbf{v}_0, \mathbf{w}_0)$ denotes its derivative with respect to $\mathbf{v}$ at the point $(\mathbf{v}_0, \mathbf{w}_0)$. For given real map $h(\mathbf{v})$ with $\mathbf{v} \in \mathbb{R}^n$, $Dh(\mathbf{v}_0)$ denotes its derivative with respect to $\mathbf{v}$ at the point $\mathbf{v}_0$. In addition, $\|\cdot\|$ represent the induced 2-norm.

2. Problem formulation

Consider the following nonlinear system:

$$\begin{cases} \dot{\mathbf{x}} = f(\mathbf{x}, \mathbf{u}) \\ \mathbf{y} = h(\mathbf{x}) \end{cases} \tag{1}$$

where $\mathbf{x} \in \mathcal{X} \subset \mathbb{R}^n$ is the state vector with $\mathcal{X}$ a open subset of $\mathbb{R}^n$, $\mathbf{y} \in \mathbb{R}^l$ is the measurement output vector, and $\mathbf{u} \in \mathbb{R}^m$ is the control input vector satisfying the constraints

$$\mathbf{u} \in \Omega \stackrel{\triangle}{=} \left\{ \mathbf{u} = [u_1 \ u_2 \ \cdots \ u_m]^T \Big| \underline{u}_i \leq u_i \leq \bar{u}_i, \ i = 1, 2, \cdots, m \right\} \tag{2}$$

with $\underline{\mathbf{u}} = [\underline{u}_1 \ \underline{u}_2 \ \cdots \ \underline{u}_m]^T$ and $\bar{\mathbf{u}} = [\bar{u}_1 \ \bar{u}_2 \ \cdots \ \bar{u}_m]^T$ being vectors of lower and upper control limits, respectively.

We assume that the system (1) satisfies the following assumption:

Assumption 1. *The function $f(\mathbf{x}, \mathbf{u})$ is smooth and the output function $h(\mathbf{x})$ is continuously differentiable.*

Since control allocation need full state information, the state estimation for the system (1) is required.

Consider a dynamic observer of the following form

$$\dot{\hat{\mathbf{x}}} = f(\hat{\mathbf{x}}, \mathbf{u}) - \Phi(\hat{\mathbf{x}}, \mathbf{u})[\mathbf{y} - h(\hat{\mathbf{x}})] \tag{3}$$

Define the error $\mathbf{e}$ as

$$\mathbf{e} = \mathbf{x} - \hat{\mathbf{x}} \tag{4}$$

To estimate the state $\mathbf{x}$, we wish to design the mapping $\Phi(\hat{\mathbf{x}}, \mathbf{u})$ such that the trajectory of $\mathbf{e}$ with the dynamics

$$\dot{\mathbf{e}} = f(\mathbf{x}, \mathbf{u}) - f(\hat{\mathbf{x}}, \mathbf{u}) + \Phi(\hat{\mathbf{x}}, \mathbf{u})[\mathbf{y} - h(\hat{\mathbf{x}})] \tag{5}$$

exponentially converges to zero as $t \to +\infty$, uniformly on $\mathbf{u} \in \Omega$, for every $\mathbf{x}(0)$ subject to $\mathbf{e}(0) = \mathbf{x}(0) - \hat{\mathbf{x}}(0)$ near zero.

The aim is to design a nonlinear control allocation law based on the state observer (3) such that a reference model that represents a predefined dynamics of the closed-loop system is tracked subject to the control constraint $\mathbf{u} \in \Omega$.

Given that the predefined dynamics of the closed-loop system is described by the following asymptotically stable reference model

$$\dot{\mathbf{x}} = \mathbf{A}_d \mathbf{x} + \mathbf{B}_d \mathbf{r} \tag{6}$$

where $\mathbf{A}_d \in \mathbb{R}^{n \times n}$, $\mathbf{B}_d \in \mathbb{R}^{n \times n_r}$ and the reference $\mathbf{r} \in \mathbb{R}^{n_r}$ satisfy the following assumption.

Assumption 2. $\mathbf{A}_d$ *is Hurwitz, and $\mathbf{r} \in \Sigma \subset \mathbb{R}^{n_r}$ is continuously differentiable where Σ is an open subset defined by: for each $\mathbf{r} \in \Sigma$, there exist $\mathbf{x} \in \mathcal{X}$ and $\mathbf{u} \in \Omega$ such that the system (1) matches the reference system (6).*

Since the state $\mathbf{x}$ is unmeasurable, the control allocation design is then based on its estimate $\hat{\mathbf{x}}$. In other words, we have to first choose the mapping $\Phi(\hat{\mathbf{x}}, \mathbf{u})$ in (3) such that the estimation error $\mathbf{e}$ exponentially converges to zero as $t \to +\infty$, uniformly on $\mathbf{u} \in \Omega$, for every $\mathbf{x}(0) \in \mathcal{X}$ subject to $\mathbf{e}(0)$ near zero; then minimize the cost function

$$J(\hat{\mathbf{x}}, \mathbf{r}, \mathbf{u}) = \frac{1}{2}\mathbf{u}^T\mathbf{H}_1\mathbf{u} + \frac{1}{2}\tau^T(\hat{\mathbf{x}}, \mathbf{r}, \mathbf{u})\mathbf{H}_2\tau(\hat{\mathbf{x}}, \mathbf{r}, \mathbf{u}) \tag{7}$$

where $\mathbf{H}_1 \in \mathbb{R}^{m\times m}$ and $\mathbf{H}_2 \in \mathbb{R}^{n\times n}$ are positive definite weighting matrices, and

$$\tau(\hat{\mathbf{x}}, \mathbf{r}, \mathbf{u}) \stackrel{\triangle}{=} f(\hat{\mathbf{x}}, \mathbf{u}) - \mathbf{A}_d\hat{\mathbf{x}} - \mathbf{B}_d\mathbf{r} \tag{8}$$

is the matching error between the actual dynamics and desired dynamics. Since power consumption minimization introduced by the term $\frac{1}{2}\mathbf{u}^T\mathbf{H}_1\mathbf{u}$ is a secondary objective, we choose $\|\mathbf{H}_1\| \ll \|\mathbf{H}_2\|$.

Now the control allocation problem is formulated in terms of solving the following nonlinear static minimization problem:

$$\begin{array}{rl} & \min_{\mathbf{u}} J(\hat{\mathbf{x}}, \mathbf{r}, \mathbf{u}) \quad \text{subject to} \\ \mathbf{u} \in \Omega & \text{and } \hat{\mathbf{x}} \text{ converges to } \mathbf{x} \text{ exponentially} \end{array} \tag{9}$$

Define

$$\Delta(\mathbf{u}) = [S(u_1) \;\; S(u_2) \;\; \cdots \;\; S(u_m)] \tag{10}$$

with

$$S(u_i) = \min((u_i - \underline{u}_i)^3, (\bar{u}_i - u_i)^3, 0), i = 1, 2, \cdots, m \tag{11}$$

Then the constraint condition $\mathbf{u} \in \Omega$ is equivalent to

$$\Delta(\mathbf{u}) = 0 \tag{12}$$

Introduce the Lagrangian

$$L(\hat{\mathbf{x}}, \mathbf{r}, \mathbf{u}, \lambda) = J(\hat{\mathbf{x}}, \mathbf{r}, \mathbf{u}) + \Delta(\mathbf{u})\lambda \tag{13}$$

where $\lambda \in \mathbb{R}^m$ is a Lagrange multiplier. And assume that

Assumption 3. *There exists a constant $\gamma_1 > 0$ such that $\frac{\partial^2 L}{\partial \mathbf{u}^2} \geq \gamma_1 \mathbf{I}_m$.*

The following lemma is immediate ((Wismer & Chattergy, 1978), p. 42).

Lemma 1. *If Assumptions 1 and 3 hold, the Lagrangian (13) achieves a local minimum if and only if $\frac{\partial L}{\partial \lambda} = 0$ and $\frac{\partial L}{\partial \mathbf{u}} = 0$.*

Proof. Necessity: The necessary condition is obvious. *Sufficiency:* Since $\frac{\partial L}{\partial \lambda} = 0$, we have $\Delta(\mathbf{u}) = 0$. In this case, the Lagrangian (13) is independent of the Lagrange multiplier λ, which achieves a local minimum if $\frac{\partial L}{\partial \mathbf{u}} = 0$ and $\frac{\partial^2 L}{\partial \mathbf{u}^2} > 0$. As $\frac{\partial^2 L}{\partial \mathbf{u}^2} > 0$ is guaranteed by Assumption 3, thus, $\frac{\partial L}{\partial \lambda} = 0$ and $\frac{\partial L}{\partial \mathbf{u}} = 0$ implies the local minimum. The proof is completed. □

Remark 1. *It should be noted that Assumption 3 is satisfied if all control inputs are within their limits (i.e., $\frac{\partial L}{\partial \lambda} = \Delta^T(\mathbf{u}) = 0$) and the nonlinear system (1) is affine in control (i.e., $f(\mathbf{x},\mathbf{u}) = f_1(\mathbf{x}) + g(\mathbf{x})\mathbf{u}$). It is because, in this case, $\frac{\partial^2 L}{\partial \mathbf{u}^2} = \mathbf{H}_1 + g^T(\mathbf{x})\mathbf{H}_2 g(\mathbf{x})$ is positive definite matrix for $\mathbf{H}_1 > 0$ and $\mathbf{H}_2 > 0$. Furthermore, since the Lagrangian (13) is convex in this case, Lemma 1 holds for a global minimum.*

To solve the control allocation problem (9) with the state estimate $\hat{\mathbf{x}}$ from the observer (3), we consider the following control Lyapunov-like function

$$V(\hat{\mathbf{x}},\mathbf{e},\mathbf{r},\mathbf{u},\lambda) = V_m(\hat{\mathbf{x}},\mathbf{r},\mathbf{u},\lambda) + \frac{1}{2}\mathbf{e}^T\mathbf{P}\mathbf{e} \tag{14}$$

where $\mathbf{P} > 0$ is a known positive-definite matrix and

$$V_m(\hat{\mathbf{x}},\mathbf{r},\mathbf{u},\lambda) = \frac{1}{2}\left[\left(\frac{\partial L}{\partial \mathbf{u}}\right)^T \frac{\partial L}{\partial \mathbf{u}} + \left(\frac{\partial L}{\partial \lambda}\right)^T \frac{\partial L}{\partial \lambda}\right] \tag{15}$$

Here the function V_m is designed to attract $(\mathbf{u},\lambda)$ so as to minimize the Lagrangian (13). The term $\frac{1}{2}\mathbf{e}^T\mathbf{P}\mathbf{e}$ forms a standard Lyapunov-like function for observer estimation error $\mathbf{e}$ which is required to exponentially converge to zero as $t \to +\infty$.

Following the observer design in (Tsinias, 1989), we define a neighborhood Q of zero with $Q \subset \mathcal{X}$, a neighborhood W of $\mathcal{X}$ with $\{\mathbf{x} - \mathbf{e} : \mathbf{x} \in \mathcal{X}, \mathbf{e} \in Q\} \subset W$, and a closed ball S of radius $r > 0$, centered at zero, such that $S \subset Q$. Then define the boundary of S as ∂S. Figure 1 illustrates the geometrical relationship of these defined sets.

Let $\mathcal{H}$ denote the set of the continuously differentiable output mappings $h(\mathbf{x}) : \mathcal{X} \to \mathbb{R}^l$ such that for every $\mathbf{m}_0 \in Q$ and $\hat{\mathbf{x}} \in W$,

$$R(\hat{\mathbf{x}},\mathbf{m}_0) \geq 0 \tag{16}$$

and

$$\ker R(\hat{\mathbf{x}},\mathbf{m}_0) \subset \ker Dh(\hat{\mathbf{x}}) \tag{17}$$

where

$$R(\hat{\mathbf{x}},\mathbf{m}_0) \triangleq [Dh(\hat{\mathbf{x}})]^T Dh(\hat{\mathbf{x}} + \mathbf{m}_0) + [Dh(\hat{\mathbf{x}} + \mathbf{m}_0)]^T Dh(\hat{\mathbf{x}}) \tag{18}$$

Remark 2. *Obviously, every linear map $\mathbf{y} = \mathbf{H}\mathbf{x}$ belongs to $\mathcal{H}$. Furthermore, $\mathcal{H}$ contains a wide family of nonlinear mappings.*

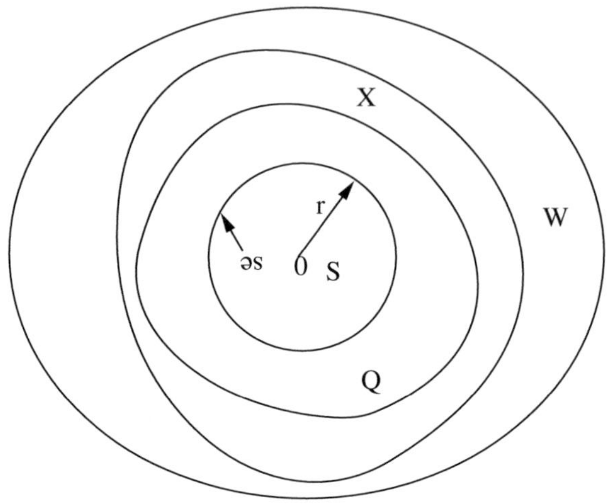

Fig. 1. Geometrical representation of sets

We assume that

Assumption 4. *$h(\mathbf{x})$ in the system (1) belongs to the set $\mathcal{H}$, namely, $h(\mathbf{x}) \in \mathcal{H}$.*

Further, we define

$$N \stackrel{\triangle}{=} \left\{ \mathbf{e} \in \mathbb{R}^{n_x} \middle| \mathbf{e}^T \mathbf{P} D_{\mathbf{x}} f(\hat{\mathbf{x}} + \mathbf{m}_1, \mathbf{u}) \mathbf{e} \leq -k_0 \|\mathbf{e}\|^2 \right\} \tag{19}$$

and assume that

Assumption 5. *There exist a positive definite matrix $\mathbf{P} \in \mathbb{R}^{n_x \times n_x}$ and a positive constant k_0 such that $\ker Dh(\hat{\mathbf{x}}) \subset N$ holds for any $(\hat{\mathbf{x}}, \mathbf{m}_1, \mathbf{u}) \in W \times Q \times \Omega$.*

Remark 3. *Assumption 5 ensures that the estimation error system (5) is stable in the case of $h(\mathbf{x}) = h(\hat{\mathbf{x}})$ and $\mathbf{x} \neq \hat{\mathbf{x}}$. In particular, for linear systems, the condition in Assumption 5 is equivalent to detectability.*

3. Main results

Denote

$$\begin{bmatrix} \alpha \\ \beta \end{bmatrix} = \begin{bmatrix} \dfrac{\partial^2 L}{\partial \mathbf{u}^2} & \dfrac{\partial^2 L}{\partial \lambda \partial \mathbf{u}} \\ \dfrac{\partial^2 L}{\partial \mathbf{u} \partial \lambda} & \mathbf{0} \end{bmatrix} \begin{bmatrix} \dfrac{\partial L}{\partial \mathbf{u}} \\ \dfrac{\partial L}{\partial \lambda} \end{bmatrix} \tag{20}$$

and define

$$M \stackrel{\triangle}{=} \left\{ \nu \in \mathbb{R}^{n_x} \middle| \ \nu = r \|\mathbf{e}\|^{-1} \mathbf{e}, \ \mathbf{e} \in N \cap S \right\} \tag{21}$$

Let

$$\gamma_1(\hat{\mathbf{x}}, \mathbf{u}) = \max \left\{ r^2 (\|\mathbf{P}\| \|D_{\mathbf{x}} f(\hat{\mathbf{x}} + \mathbf{m}_1, \mathbf{u})\| + k_0), \ \mathbf{m}_1 \in S, \ (\hat{\mathbf{x}}, \mathbf{u}) \in W \times \Omega \right\} \tag{22}$$

$$\gamma_2(\hat{\mathbf{x}}) = \min \left\{ \frac{1}{2} \nu^T R(\hat{\mathbf{x}}, \mathbf{m}_0) \nu, \ \mathbf{m}_0 \in S, \ \nu \in \partial S - M, \ \hat{\mathbf{x}} \in W \right\} \tag{23}$$

Theorem 1. *Consider the system (1) with $\mathbf{x} \in \mathcal{X}$ and $\mathbf{u} \in \Omega$. Suppose that Assumptions 1-5 are satisfied. For a given asymptotically stable matrix $\mathbf{A}_d$ and a matrix $\mathbf{B}_d$, given symmetric positive-definite matrices Γ_1 and Γ_2, and a given positive constants ω, for $\mathbf{e}(0)$ near zero, $\left(\frac{\partial L}{\partial \lambda}, \frac{\partial L}{\partial \mathbf{u}}, \mathbf{e}\right)$ exponentially converges to zero as $t \to +\infty$, and the dynamics of the nonlinear system (1) exponentially converges to that of the stable system (6) if the following dynamic update law*

$$\begin{cases} \dot{\mathbf{u}} = -\Gamma_1 \alpha + \xi_1 \\ \dot{\lambda} = -\Gamma_2 \beta + \xi_2 \end{cases} \tag{24}$$

and the observer system

$$\dot{\hat{\mathbf{x}}} = f(\hat{\mathbf{x}}, \mathbf{u}) - \Phi(\hat{\mathbf{x}}, \mathbf{u}) \left[\mathbf{y} - h(\hat{\mathbf{x}})\right] \tag{25}$$

are adopted. Here $\alpha, \beta \in \mathbb{R}^m$ are as in (20), and $\xi_1, \xi_2 \in \mathbb{R}^m$ satisfy

$$\alpha^T \xi_1 + \beta^T \xi_2 + \delta + \omega V_m = 0 \tag{26}$$

with V_m as in (15) and

$$\delta = \left(\frac{\partial L}{\partial \mathbf{u}}\right)^T \frac{\partial^2 L}{\partial \mathbf{r} \partial \mathbf{u}} \dot{\mathbf{r}} + \left(\frac{\partial L}{\partial \mathbf{u}}\right)^T \frac{\partial^2 L}{\partial \hat{\mathbf{x}} \partial \mathbf{u}} \dot{\hat{\mathbf{x}}} \tag{27}$$

and the mapping

$$\Phi(\hat{\mathbf{x}}, \mathbf{u}) = -\theta(\hat{\mathbf{x}}, \mathbf{u}) \mathbf{P}^{-1} [Dh(\hat{\mathbf{x}})]^T \tag{28}$$

where

$$\theta(\hat{\mathbf{x}}, \mathbf{u}) \geq \frac{\gamma_1(\hat{\mathbf{x}}, \mathbf{u})}{\gamma_2(\hat{\mathbf{x}})} > 0 \tag{29}$$

with $\gamma_1(\hat{\mathbf{x}}, \mathbf{u}) > 0$ and $\gamma_2(\hat{\mathbf{x}}) > 0$ defined as in (22) and (23).

Proof. From the Lyapunov-like function (14), we obtain its time derivative as

$$\dot{V} = \left[\left(\frac{\partial L}{\partial \mathbf{u}}\right)^T \frac{\partial^2 L}{\partial \mathbf{u}^2} + \left(\frac{\partial L}{\partial \lambda}\right)^T \frac{\partial^2 L}{\partial \mathbf{u} \partial \lambda}\right] \dot{\mathbf{u}} + \left(\frac{\partial L}{\partial \mathbf{u}}\right)^T \frac{\partial^2 L}{\partial \lambda \partial \mathbf{u}} \dot{\lambda}$$
$$+ \left(\frac{\partial L}{\partial \mathbf{u}}\right)^T \frac{\partial^2 L}{\partial \mathbf{r} \partial \mathbf{u}} \dot{\mathbf{r}} + \left(\frac{\partial L}{\partial \mathbf{u}}\right)^T \frac{\partial^2 L}{\partial \hat{\mathbf{x}} \partial \mathbf{u}} \dot{\hat{\mathbf{x}}} + \mathbf{e}^T \mathbf{P} \dot{\mathbf{e}} \tag{30}$$

Substituting $\dot{\mathbf{e}}$ in (5), α and β as in (20) and δ as in (27) into (30), we have

$$\dot{V} = \alpha^T \dot{\mathbf{u}} + \beta^T \dot{\lambda} + \delta + \mathbf{e}^T \mathbf{P} \left\{ f(\mathbf{x}, \mathbf{u}) - f(\hat{\mathbf{x}}, \mathbf{u}) + \Phi(\hat{\mathbf{x}}, \mathbf{u}) [\mathbf{y} - h(\hat{\mathbf{x}})] \right\} \tag{31}$$

Consider $\mathbf{e} \in S$. Since S is convex, according to Mean Value Theorem, there exists $\mathbf{m}_0, \mathbf{m}_1 \in S$ satisfying

$$f(\mathbf{x}, \mathbf{u}) - f(\hat{\mathbf{x}}, \mathbf{u}) = D_{\mathbf{x}} f(\hat{\mathbf{x}} + \mathbf{m}_1, \mathbf{u}) \mathbf{e} \tag{32}$$
$$\mathbf{y} - h(\hat{\mathbf{x}}) = Dh(\hat{\mathbf{x}} + \mathbf{m}_0) \mathbf{e} \tag{33}$$

Then substituting (24), (26), (32) and (33) into (31), we obtain

$$\dot{V} = -\alpha^T \Gamma_1 \alpha - \beta^T \Gamma_2 \beta - \omega V_m + \mathbf{e}^T \mathbf{P} \left[D_\mathbf{x} f(\hat{\mathbf{x}} + \mathbf{m}_1, \mathbf{u}) + \Phi(\hat{\mathbf{x}}, \mathbf{u}) D h(\hat{\mathbf{x}} + \mathbf{m}_0) \right] \mathbf{e} \tag{34}$$

After substituting $\Phi(\hat{\mathbf{x}}, \mathbf{u})$ as in (28) and $R(\hat{\mathbf{x}}, m_0)$ as in (18), (34) can be rewritten as

$$\dot{V} = -\alpha^T \Gamma_1 \alpha - \beta^T \Gamma_2 \beta - \omega V_m + \mathbf{e}^T \mathbf{P} D_\mathbf{x} f(\hat{\mathbf{x}} + \mathbf{m}_1, \mathbf{u}) \mathbf{e} - \frac{\theta(\hat{\mathbf{x}}, \mathbf{u})}{2} \mathbf{e}^T R(\hat{\mathbf{x}}, \mathbf{m}_0) \mathbf{e} \tag{35}$$

Since the matrices $\Gamma_1 > 0$ and $\Gamma_2 > 0$, we have

$$\dot{V} \le -\omega V_m + \mathbf{e}^T \mathbf{P} D_\mathbf{x} f(\hat{\mathbf{x}} + \mathbf{m}_1, \mathbf{u}) \mathbf{e} - \frac{\theta(\hat{\mathbf{x}}, \mathbf{u})}{2} \mathbf{e}^T R(\hat{\mathbf{x}}, \mathbf{m}_0) \mathbf{e} \tag{36}$$

For $\mathbf{e} = 0$ where $\mathbf{x}$ is determined by the observer accurately, we have

$$\dot{V} \le -\omega V_m = -\omega V \tag{37}$$

Since $\omega > 0$, V exponentially converges to zero as $t \to +\infty$. Hence, $\left(\frac{\partial L}{\partial \lambda}, \frac{\partial L}{\partial \mathbf{u}} \right)$ exponentially converges to zero.

For any nonzero $\mathbf{e} \in S$, let $\nu = r \|\mathbf{e}\|^{-1} \mathbf{e}$. Obviously, $\nu \in \partial S$. Then we have

$$\dot{V} \le -\omega V_m + \frac{1}{r^2} \|\mathbf{e}\|^2 \nu^T \mathbf{P} D_\mathbf{x} f(\hat{\mathbf{x}} + \mathbf{m}_1, \mathbf{u}) \nu - \frac{\theta(\hat{\mathbf{x}}, \mathbf{u})}{2r^2} \|\mathbf{e}\|^2 \nu^T R(\hat{\mathbf{x}}, \mathbf{m}_0) \nu \tag{38}$$

In the following, we shall show that V converges exponentially to zero for all $\mathbf{m}_0$, $\mathbf{m}_1 \in S$, $\hat{\mathbf{x}} \in W$, $\mathbf{u} \in \Omega$, $\mathbf{e} \in S$, $\mathbf{e} \ne 0$ and $\nu \in \partial S$.

First let us consider nonzero $\mathbf{e} \in N \cap S$. From $\nu = r \|\mathbf{e}\|^{-1} \mathbf{e}$, we have $\nu \in M$. Since $\mathbf{m}_0$, $\mathbf{m}_1 \in S \subset Q$, $\hat{\mathbf{x}} \in W$ and $\mathbf{u} \in \Omega$, according to Assumptions 1-5, it follows that

$$\nu^T R(\hat{\mathbf{x}}, \mathbf{m}_0) \nu = 0 \tag{39}$$

and

$$\nu^T \mathbf{P} D_\mathbf{x} f(\hat{\mathbf{x}} + \mathbf{m}_1, \mathbf{u}) \nu \le -k_0 \|\nu\|^2 \tag{40}$$

with the constant $k_0 > 0$. From (38), we have

$$\dot{V} \le -\omega V_m - k_0 \|\mathbf{e}\|^2 \le -\sigma V \tag{41}$$

with the constant $\sigma > 0$. Hence, $\left(\frac{\partial L}{\partial \lambda}, \frac{\partial L}{\partial \mathbf{u}}, \mathbf{e} \right)$ exponentially converges to zero as $t \to +\infty$.

Then we consider nonzero $\mathbf{e} \in S - N \cap S$, namely, $\nu \in \partial S - M$. From (38), taking into account (22)-(23), we obtain

$$\dot{V} \le -\omega V_m + \frac{1}{r^2} \|\mathbf{e}\|^2 \left[\gamma_1(\hat{\mathbf{x}}, \mathbf{u}) - k_0 r^2 - \theta(\hat{\mathbf{x}}, \mathbf{u}) \gamma_2(\hat{\mathbf{x}}) \right] \tag{42}$$

Since $\theta(\hat{\mathbf{x}}, \mathbf{u})$ satisfy the condition (29), we obtain (41) again. Hence, in this case, $\left(\frac{\partial L}{\partial \lambda}, \frac{\partial L}{\partial \mathbf{u}}, \mathbf{e} \right)$ also exponentially converges to zero as $t \to +\infty$.

Since $\left(\frac{\partial L}{\partial \lambda}, \frac{\partial L}{\partial \mathbf{u}}, \mathbf{e}\right)$ exponentially converges to zero as $t \rightarrow +\infty$, the closed-loop system exponentially converges to

$$\begin{cases} \dot{\hat{\mathbf{x}}} = \mathbf{A}_d\hat{\mathbf{x}} + \mathbf{B}_d\mathbf{r} \\ \dot{\mathbf{e}} = \left\{D_{\mathbf{x}}f(\hat{\mathbf{x}} + \mathbf{m}_1, \mathbf{u}) - \theta(\hat{\mathbf{x}}, \mathbf{u})\mathbf{P}^{-1}[Dh(\hat{\mathbf{x}})]^T Dh(\hat{\mathbf{x}} + \mathbf{m}_0)\right\}\mathbf{e} \end{cases} \tag{43}$$

Since $\mathbf{A}_d$ is a asymptotically stable matrix, we know that $\hat{\mathbf{x}} \in W$ is bounded. According to Assumptions 1 and 4, $D_{\mathbf{x}}f(\hat{\mathbf{x}} + \mathbf{m}_1, \mathbf{u})$, $Dh(\hat{\mathbf{x}})$ and $Dh(\hat{\mathbf{x}} + \mathbf{m}_0)$ are all bounded for $\mathbf{m}_0, \mathbf{m}_1 \in S$ and $\mathbf{u} \in \Omega$. From $k_0 > 0$, we have $0 < \gamma_1(\hat{\mathbf{x}}, \mathbf{u}) < +\infty$. According to Assumption 4, we have $\ker R(\hat{\mathbf{x}}, \mathbf{m}_0) \subset \ker Dh(\hat{\mathbf{x}})$ which ensures that $0 < \nu^T R(\hat{\mathbf{x}}, \mathbf{m}_0)\nu < +\infty$ for every $\nu \in \partial S - M$, $\mathbf{m}_0 \in S$ and $\hat{\mathbf{x}} \in W$. Thus, we have $0 < \gamma_2(\hat{\mathbf{x}}) < +\infty$. As a result, $0 < \theta(\hat{\mathbf{x}}, \mathbf{u}) < +\infty$. From (43), we know that $\dot{\mathbf{e}}$ exponentially converges to zero as $\mathbf{e}$ exponentially converges to zero. Moreover, we have

$$\dot{\mathbf{x}} - \dot{\mathbf{e}} = \mathbf{A}_d\mathbf{x} - \mathbf{A}_d\mathbf{e} + \mathbf{B}_d\mathbf{r} \tag{44}$$

Since $\dot{\mathbf{e}}$ and $\mathbf{e}$ exponentially converges to zero, we have the system (1) exponentially converges to $\dot{\mathbf{x}} = \mathbf{A}_d\mathbf{x} + \mathbf{B}_d\mathbf{r}$. This completes the proof. □

Consider now the issue of solving (26) with respect to ξ_1 and ξ_2. One method to achieve a well-defined unique solution to the under-determined algebraic equation is to solve a least-square problem subject to (26). This leads to the Lagrangian

$$l(\xi_1, \xi_2, \rho) = \frac{1}{2}(\xi_1^T\xi_1 + \xi_2^T\xi_2) + \rho(\alpha^T\xi_1 + \beta^T\xi_2 + \delta + \omega V_m) \tag{45}$$

where $\rho \in \mathbb{R}$ is a Lagrange multiplier. The first order optimality conditions

$$\frac{\partial l}{\partial \xi_1} = 0, \quad \frac{\partial l}{\partial \xi_2} = 0, \quad \frac{\partial l}{\partial \rho} = 0 \tag{46}$$

leads to the following system of linear equations

$$\begin{bmatrix} I_m & 0 & \alpha \\ 0 & I_m & \beta \\ \alpha^T & \beta^T & 0 \end{bmatrix} \begin{bmatrix} \xi_1 \\ \xi_2 \\ \rho \end{bmatrix} = \begin{bmatrix} 0 \\ 0 \\ -\delta - \omega V_m \end{bmatrix} \tag{47}$$

Remark 4. *It is noted that Equation (47) always has a unique solution for ξ_1 and ξ_2 if any one of α and β is nonzero.*

4. Example

Consider the pendulum system

$$\begin{bmatrix} \dot{x}_1 \\ \dot{x}_2 \end{bmatrix} = \begin{bmatrix} x_2 \\ -\sin x_1 + u_1 \cos x_1 + u_2 \sin x_1 \end{bmatrix} \tag{48}$$

$$y = x_1 + x_2 \tag{49}$$

with $\mathbf{x} = [x_1 \;\; x_2]^T \in \mathbb{R}^2$, $\mathbf{u} = [u_1 \;\; u_2]^T \in \Omega$ and

$$\Omega \triangleq \left\{ \mathbf{u} = [u_1 \;\; u_2]^T \Big| -1 \le u_1 \le 1, \;\; -0.5 \le u_2 \le 0.5 \right\} \tag{50}$$

As the system is affine in control and its measurement output y is a linear map of its state $\mathbf{x}$, Assumptions 1, 3 and 4 are satisfied automatically.

Choose

$$\mathbf{P} = \begin{bmatrix} 3 & 0 \\ 0 & 1 \end{bmatrix}$$

For $\mathbf{e} \neq 0$ and $\mathbf{e} \in \ker[1 \;\; 1]$, we have $e_1 = -e_2$ and

$$\begin{aligned}
&\mathbf{e}^T \mathbf{P} D_{\mathbf{x}} f(\mathbf{x}, u)\mathbf{e}|_{e_1=-e_2} \\
&= [e_1 \;\; e_2] \begin{bmatrix} 0 & 3 \\ -\cos x_1 - u_1 \sin x_1 + u_2 \cos x_1 & 0 \end{bmatrix} \begin{bmatrix} e_1 \\ e_2 \end{bmatrix} |_{e_1=-e_2} \\
&= (-\cos x_1 - u_1 \sin x_1 + u_2 \cos x_1 + 3) e_1 e_2 |_{e_1=-e_2} \\
&\le [-1.5 \cos(\arctan \tfrac{2}{3}) - \sin(\arctan \tfrac{2}{3}) + 3] e_1 e_2 |_{e_1=-e_2} \\
&= (-1.8028 + 3) e_1 e_2 |_{e_1=-e_2} \\
&= -0.5986 \|\mathbf{e}\|^2 |_{e_1=-e_2} \\
&< -k_0 \|\mathbf{e}\|^2 |_{e_1=-e_2}
\end{aligned}$$

with $0 < k_0 < 0.5986$. Hence, Assumption 5 is satisfied. Let S be the ball of radius $r = 1$, centered at zero and ∂S is the boundary of S. Define $M \subset \partial S$ and

$$M = \left\{ \nu = [\nu_1 \;\; \nu_2]^T \in \mathbb{R}^2 : \|\nu\| = 1, \; 3\nu_1\nu_2 + 1.8028|\nu_1\nu_2| < -k_0 \right\}$$

Obviously,

$$\partial S - M = \left\{ \nu = [\nu_1 \;\; \nu_2]^T \in \mathbb{R}^2 : \|\nu\| = 1, \; 3\nu_1\nu_2 + 1.8028|\nu_1\nu_2| \ge -k_0 \right\}$$

As $\gamma_1(\hat{\mathbf{x}}, u) = 3 \times 1.8028 + k_0$ and

$$\gamma_2(\hat{\mathbf{x}}) = \min \left\{ (\nu_1 + \nu_2)^2, \;\; \nu \in \partial S - M \right\} = 1 - \frac{2k_0}{3 - 1.8028}$$

choosing $k_0 = 0.5$, we have $\dfrac{\gamma_1(\hat{\mathbf{x}}, \mathbf{u})}{\gamma_2(\hat{\mathbf{x}})} = 35.8699$. Let $\theta(\hat{\mathbf{x}}, \mathbf{u}) = 36 > 35.8699$ and we have $\Phi(\hat{\mathbf{x}}, \mathbf{u}) = -[12 \;\; 36]^T$.

Now the nonlinear observer becomes

$$\begin{bmatrix} \dot{\hat{x}}_1 \\ \dot{\hat{x}}_2 \end{bmatrix} = \begin{bmatrix} \hat{x}_2 \\ -\sin \hat{x}_1 + u_1 \cos \hat{x}_1 + u_2 \sin \hat{x}_1 \end{bmatrix} + \begin{bmatrix} 12 \\ 36 \end{bmatrix} (y - \hat{x}_1 - \hat{x}_2)$$

Choose the reference model (6) where

$$\mathbf{A}_d = \begin{bmatrix} 0 & 1 \\ -25 & -10 \end{bmatrix}, \quad \mathbf{B}_d = \begin{bmatrix} 0 \\ 25 \end{bmatrix}$$

and the reference is given by

$$\mathbf{r} = \begin{cases} r_f \left[6\left(\frac{t}{t_1}\right)^5 - 15\left(\frac{t}{t_1}\right)^4 + 10\left(\frac{t}{t_1}\right)^3 \right], & 0 \le t < t_1 \\ r_f, & t_1 \le t < t_2 \\ -r_f \left[6\left(\frac{t-t_2}{t_f-t_2}\right)^5 - 15\left(\frac{t-t_2}{t_f-t_2}\right)^4 + 10\left(\frac{t-t_2}{t_f-t_2}\right)^3 \right] + r_f, & t_2 \le t < t_f \\ 0, & t \ge t_f \end{cases}$$

with $t_1 = 10s$, $t_2 = 20s$, $t_f = 30s$ and $r_f = 0.5$. Obviously, Assumption 2 is satisfied.

Set $\mathbf{H}_1 = 0$, $\mathbf{H}_2 = 10^{-4}\mathbf{I}_2$, $\omega = 1$, $\Gamma_1 = \Gamma_2 = 2\mathbf{I}_2$, and $x_1(0) = 0.3$ and $x_2(0) = 0.5$. Using the proposed approach, we have the simulation result of the pendulum system (48)-(50) shown in Figures 2-5 where the control u_2 is stuck at -0.5 from $t = 12s$ onward.

From Figure 2, it is observed that the estimated states $\hat{x}_1$ and $\hat{x}_2$ converge to the actual states x_1 and x_2 and match the desired states x_{1d} and x_{2d} well, respectively, even when u_2 is stuck at -0.5. This observation is further verified by Figure 3 where both the state estimation errors $e_1(= x_1 - \hat{x}_1)$ and $e_2(= x_2 - \hat{x}_2)$ of the nonlinear observer as in (4) and the matching errors $\tau_1(= 0)$ and $\tau_2(= -\sin\hat{x}_1 + u_1\cos\hat{x}_1 + u_2\sin\hat{x}_1 + 25\hat{x}_1 + 10\hat{x}_2 - 25\mathbf{r})$ as in (8) exponentially converge to zero. Moreover, Figure 4 shows that the control u_1 roughly satisfies the control constraint $u_1 \in [-1, 1]$ while the control u_2 strictly satisfies the control constraint $u_2 \in [-0.5, 0.5]$. This is because, in this example, the Lagrange multiplier λ_1 is first activated by the control $u_1 < -1$ at $t = 0$ (see Figure 5 where λ_1 is no longer zero from $t = 0$),

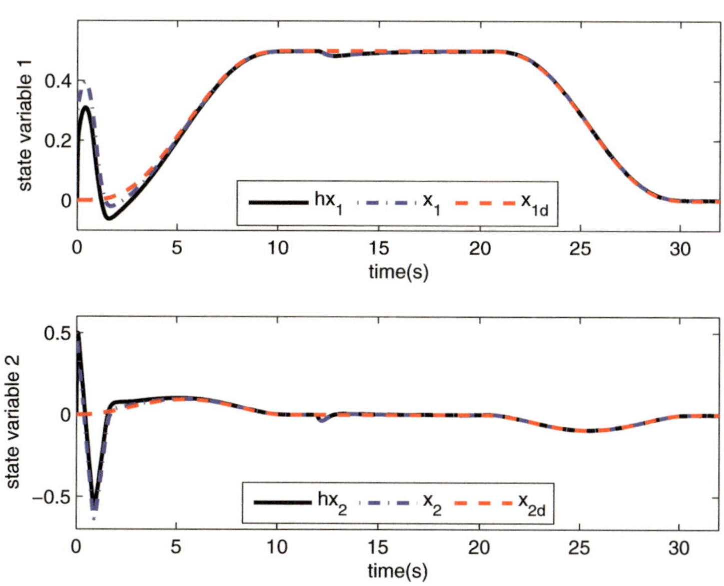

Fig. 2. Responses of the desired, estimated and actual states

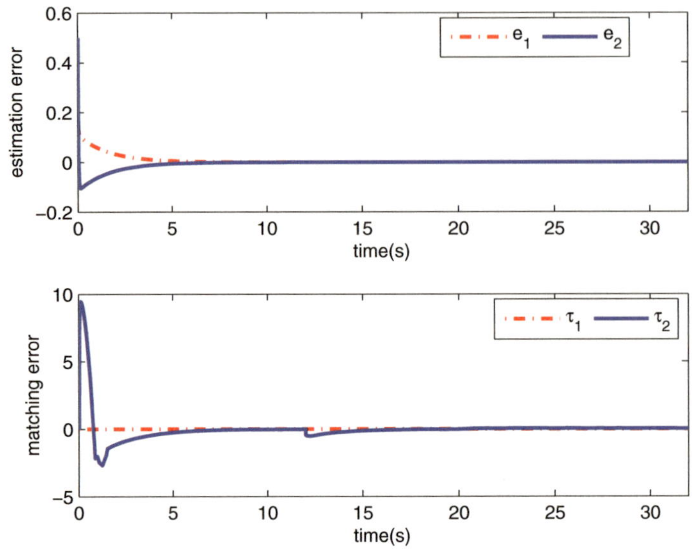

Fig. 3. Responses of estimation error and matching error

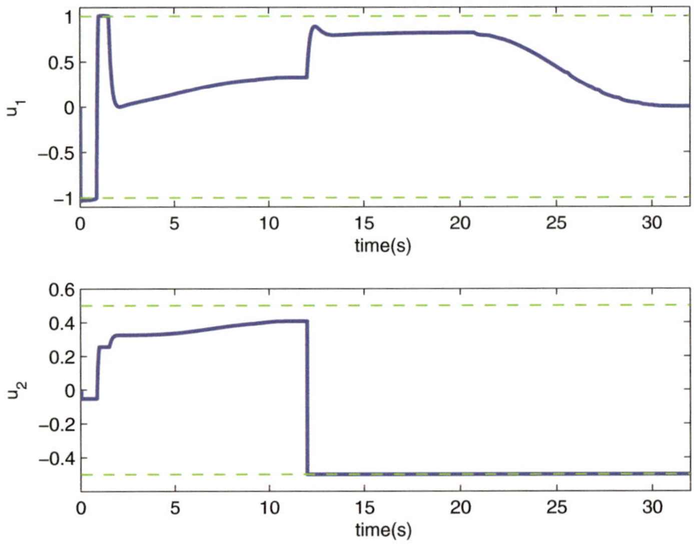

Fig. 4. Responses of control **u**

and then the proposed dynamic update law forces the control u_1 to satisfy the constraint $u_1 \in [-1, 1]$. It is also noted from Figure 5 that the Lagrange multiplier λ_2 is not activated in this example as the control u_2 is never beyond the range $[-0.5, 0.5]$. In addition, the output y and the Lyapunov-like function V_m are shown in Figure 6. From Figure 6, it is observed that the Lyapunov-like function V_m exponentially converges to zero.

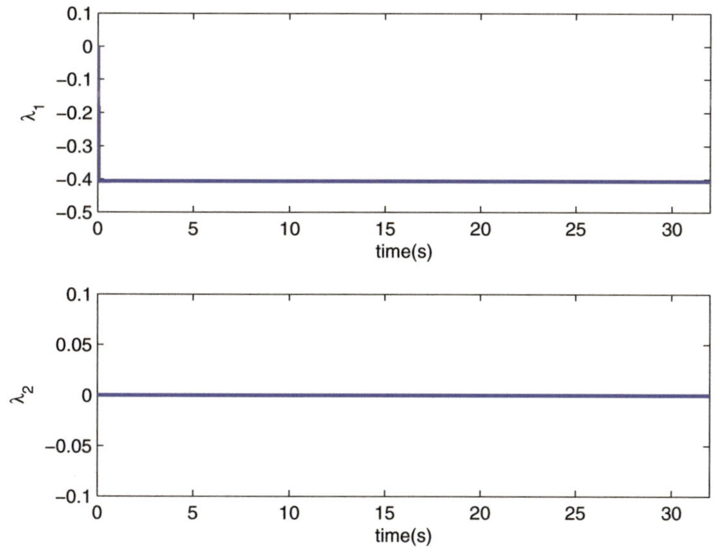

Fig. 5. Responses of Lagrangian multiplier λ

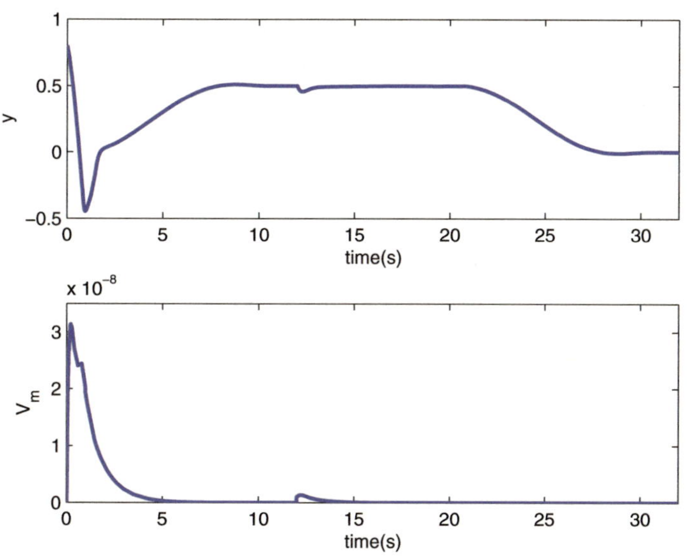

Fig. 6. Responses of output y and Lyapunov-like function V_m

5. Conclusions

Sufficient Lyapunov-like conditions have been proposed for the control allocation design via output feedback. The proposed approach is applicable to a wide class of nonlinear systems. As the initial estimation error $\mathbf{e}(0)$ need be near zero and the predefined dynamics of the closed-loop is described by a linear stable reference model, the proposed approach will present a local nature.

6. References

Ahmed-Ali, T. & Lamnabhi-Lagarrigue, F. (1999). Sliding observer-controller design for uncertain triangle nonlinear systems, *IEEE Transactions on Automatic Control* 44(6): 1244–1249.

Alamir, M. (1999). Optimization-based nonlinear observer revisited, *International Journal of Control* 72(13): 1204–1217.

Benosman, M., Liao, F., Lum, K. Y. & Wang, J. L. (2009). Nonlinear control allocation for non-minimum phase systems, *IEEE Transactions on Control Systems Technology* 17(2): 394–404.

Besancon, G. (ed.) (2007). *Nonlinear Observers and Applications*, Springer.

Besancon, G. & Ticlea, A. (2007). An immersion-based observer design for rank-observable nonlinear systems, *IEEE Transactions on Automatic Control* 52(1): 83–88.

Bestle, D. & Zeitz, M. (1983). Canonical form observer design for non-linear time-variable systems, *International Journal of Control* 38: 419–431.

Bodson, M. (2002). Evaluation of optimization methods for control allocation, *Journal of Guidance, Control and Dynamics* 25(4): 703–711.

Bornard, G. & Hammouri, H. (1991). A high gain observer for a class of uniformly observable systems, *Proceedings of the 30th IEEE Conference on Decision and Control*, pp. 1494–1496.

Buffington, J. M., Enns, D. F. & Teel, A. R. (1998). Control allocation and zero dynamics, *Journal of Guidance, Control and Dynamics* 21(3): 458–464.

Gauthier, J. P. & Kupka, I. A. K. (1994). Observability and observers for nonlinear systems, *SIAM Journal on Control and Optimization* 32: 975–994.

Krener, A. J. & Isidori, A. (1983). Linearization by output injection and nonlinear observers, *Systems & Control Letters* 3: 47–52.

Krener, A. J. & Respondek, W. (1985). Nonlinear observers with linearizable error dynamics, *SIAM Journal on Control and Optimization* 23: 197–216.

Liao, F., Lum, K. Y., Wang, J. L. & Benosman, M. (2007). Constrained nonlinear finite-time control allocation, *Proc. of 2007 American Control Conference*, New York City, NY.

Liao, F., Lum, K. Y., Wang, J. L. & Benosman, M. (2010). Adaptive Control Allocation for Non-linear Systems with Internal Dynamics, *IET Control Theory & Applications* 4(6): 909–922.

Luenberger, D. G. (1964). Observing the state of a linear system, *IEEE Transactions on Military Electron* 8: 74–80.

Michalska, H. & Mayne, D. Q. (1995). Moving horizon observers and observer-based control, *IEEE Transactions on Aotomatic Control* 40(6).

Nijmeijer, H. & Fossen, T. I. (eds) (1999). *New Directions in Nonlinear Observer Design*, Springer.

Teel, A. & Praly, L. (1994). Global stabilizability and observability imply semi-global stabilizability by output feedback, *Systems & Control Letters* 22: 313–325.

Tsinias, J. (1989). Observer design for nonlinear systems, *Systems & Control Letters* 13: 135–142.

Tsinias, J. (1990). Further results on the observer design problem, *Systems & Control Letters* 14: 411–418.

Wismer, D. A. & Chattergy, R. (1978). *Introduction To Nonlinear Optimization: A Problem Solving Approach*, Elsevier North-Holland, Inc.

Zimmer, G. (1994). State observation by on-line minimization, *International Journal of Control* 60(4): 595–606.

8

Predictive Function Control of the Single-Link Manipulator with Flexible Joint

Zhihuan Zhang and Chao Hu
Ningbo Institute of Technology, Zhejiang University
China

1. Introduction

Flexible-link robotic manipulators have many advantages with respect to conventional rigid robots. They are built by lighter, cheaper materials, which improve the payload to arm weight ratio, thus resulting in an increase of the speed with lower energy consumption. Moreover, due to the reduced inertia and compliant structure, these lightweight arms can be operated more safely and are more applicable for the delicate assembly tasks and interaction with fragile objects, including human beings.

The control for robot manipulators is to determine the time history of joint inputs to cause the end-effector to execute a commanded motion. There are many control techniques and methodologies that can be applied to the control of the manipulators. The specific control method and its implementation ways can have a significant impact on the performance of the manipulator and consequently on the range of its possible applications. In addition, the mechanical design of the manipulator itself will influence the type of control scheme needed. However, in order to improve the control performance, more sophisticated approaches should be found.

The control for flexible joint system has attracted a considerable amount of attention during the past few years. There are PD, inverse dynamics, and the force control approach for the feedback control strategies of flexible joint manipulator. (1989, MARK W. SPONG), an integral manifold approach to the feedback control of flexible joint robots (1987, MARK W. SPONG, KHASHAYAR KHORASANI, and PETAR V. KOKOTOVIC), and the nonlinear feedback control of flexible joint manipulators: a single link case study, (1990, K. KHORASANI). The basic idea of feedback linearization is to construct a nonlinear control law as a so-called inner loop control which, in the ideal case, exactly linearizes the nonlinear system after a suitable state space change of coordinates. The designer can design a second stage or outer loop control in the new coordinates to satisfy the traditional control design specifications such as tracking, disturbance rejection, and so forth. Since the feedback linearization of flexible joint manipulator is a fourth order integrator system, so we proposed a three stage design method, the first is nonlinear feedback to get integrator system, the second is pole placement to get expect performance, and the third is to use PFC to reject disturbance and uncertainty, since they can not be exactly cancelled by nonlinear feedback, coupling effects of the joint flexibility. More accurate description of robot dynamics may include fast actuator dynamics and joint-link flexibility, and so on.

2. Equations of motion

Consider the single-link arm shown in Figure 1 consisting of a flexible joint.

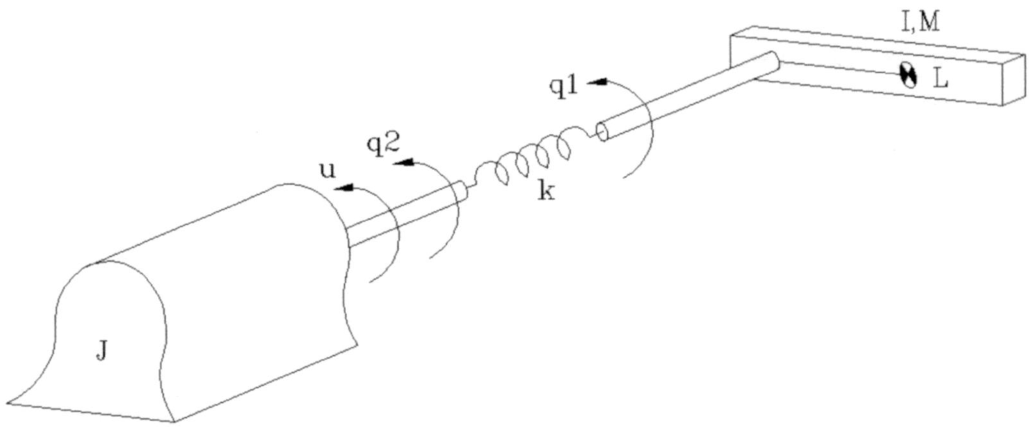

Fig. 1. Single-link robot with joint flexibility

The kinetic energy of the manipulator is a quadratic function of the vector $\dot{q}$

$$K = \frac{1}{2}\dot{q}^T D(q)\dot{q} = \frac{1}{2}\sum_{i,j}^{n} d_{ij}(q)\dot{q}_i\dot{q}_j \tag{1}$$

where the $n \times n$ inertia matrix $D(q)$ is symmetric and positive definite for each $q \in \Re^n$.

The potential energy $V = V(q)$ is independent of $\dot{q}$. We have remarked that robotic manipulator satisfies this condition.

The Euler-Lagrange equations for such a system can be derived as follows. Since

$$L = K - V = \frac{1}{2}\sum_{i,j}^{n} d_{ij}(q)\dot{q}_i\dot{q}_j - V(q) \tag{2}$$

we have

$$\frac{\partial L}{\partial \dot{q}_k} = \sum_j d_{kj}(q)\dot{q}_j$$

and

$$\frac{d}{dt}\frac{\partial L}{\partial \dot{q}_k} = \sum_j d_{kj}(q)\ddot{q}_j + \sum_j \frac{d}{dt} d_{kj}(q)\dot{q}_j = \sum_j d_{kj}(q)\ddot{q}_j + \sum_{i,j}\frac{\partial d_{kj}}{\partial q_i}\dot{q}_i\dot{q}_j$$

Also

$$\frac{\partial L}{\partial q_k} = \frac{1}{2}\sum_{i,j}\frac{\partial d_{ij}}{\partial q_k}\dot{q}_i\dot{q}_j - \frac{\partial V}{\partial q_k}$$

Thus the Euler-Lagrange equations can be written as

$$\sum_j d_{kj}(q)\ddot{q}_j + \sum_{i,j}\left\{\frac{\partial d_{kj}}{\partial q_i} - \frac{1}{2}\frac{\partial d_{ij}}{\partial q_k}\right\}\dot{q}_i\dot{q}_j - \frac{\partial V}{\partial q_k} = \tau_k\,,\ k = 1,\cdots,n \tag{3}$$

By interchanging the order of summation and taking advantage of symmetry, we can show that

$$\sum_{i,j}\left\{\frac{\partial d_{kj}}{\partial q_i}\right\}\dot{q}_i\dot{q}_j = \frac{1}{2}\sum_{i,j}\left\{\frac{\partial d_{kj}}{\partial q_i} + \frac{\partial d_{ki}}{\partial q_j}\right\}\dot{q}_i\dot{q}_j$$

Hence

$$\sum_{i,j}\left\{\frac{\partial d_{kj}}{\partial q_i} - \frac{1}{2}\frac{\partial d_{ij}}{\partial q_k}\right\}\dot{q}_i\dot{q}_j = \sum_{i,j}\frac{1}{2}\left\{\frac{\partial d_{kj}}{\partial q_i} + \frac{\partial d_{ki}}{\partial q_j} - \frac{\partial d_{ij}}{\partial q_k}\right\}\dot{q}_i\dot{q}_j$$

The term

$$c_{ijk} = \frac{1}{2}\left\{\frac{\partial d_{kj}}{\partial q_i} + \frac{\partial d_{ki}}{\partial q_j} - \frac{\partial d_{ij}}{\partial q_k}\right\} \tag{4}$$

are known as Christoffel symbols. Note that, for a fixed k, we have $c_{ijk} = c_{jik}$, which reduces the effort involved in computing these symbols by a factor of about one half. Finally, if we define

$$\varphi_k = \frac{\partial V}{\partial q_k} \tag{5}$$

then the Euler-Lagrange equations can be written as

$$\sum_j d_{kj}(q)\ddot{q}_j + \sum_{i,j} c_{ijk}(q)\dot{q}_i\dot{q}_j + \varphi_k(q) = \tau_k\,,\ k = 1,\cdots,n \tag{6}$$

In the above equation, there are three types of terms. The first involve the second derivative of the generalized coordinates. The second are quadratic terms in the first derivatives of q, where the coefficients may depend on q. These are further classified into two types. Terms involving a product of the type $\dot{q}_i^2$ are called centrifugal, while those involving a product of the type $\dot{q}_i\dot{q}_j$ where $i \neq j$ are called Coriolis terms. The third type of terms are those involving only q but not its derivatives. Clearly the latter arise from differentiating the potential energy. It is common to write (6) in matrix form as

$$D(q)\ddot{q} + C(q,\dot{q})\dot{q} + g(q) = \tau \tag{7}$$

where the k, j-th element of the matrix $C(q,\dot{q})$ is defined as

$$c_{kj} = \sum_{i=1}^{n} c_{ijk}(q)\dot{q}_i = \sum_{i=1}^{n} \frac{1}{2}\left\{\frac{\partial d_{kj}}{\partial q_i} + \frac{\partial d_{ki}}{\partial q_j} - \frac{\partial d_{ij}}{\partial q_k}\right\}\dot{q}_i$$

3. Feedback linearization design for inner loop

We first derive a model similar to (6) to represent the dynamics of a single link robot with joint flexibility. For simplicity, ignoring damping of the equations of motion, system is given by

$$D(q_1)\ddot{q}_1 + h(q_1,\dot{q}_1)\dot{q}_1 + K(q_1 - q_2) = 0 \tag{8}$$

$$J\ddot{q}_2 - K(q_1 - q_2) = u \tag{9}$$

In state space, which is now $\Re^{4n}$, we define state variables in block form

$$\begin{aligned} x_1 &= q_1 \qquad x_2 = \dot{q}_1 \\ x_3 &= q_2 \qquad x_4 = \dot{q}_2 \end{aligned} \tag{10}$$

Then from (8)-(9) we have

$$\dot{x}_1 = x_2 \tag{11}$$

$$\dot{x}_2 = -D(x_1)^{-1}\{h(x_1,x_2) + K(x_1 - x_3)\} \tag{12}$$

$$\dot{x}_3 = x_4 \tag{13}$$

$$\dot{x}_4 = J^{-1}K(x_1 - x_3) + J^{-1}u \tag{14}$$

This system is then of the form

$$\dot{x} = f(x) + G(x)u \tag{15}$$

For a single-input nonlinear system, $f(x)$ and $g(x)$ are smooth vector fields on $\Re^n$, $f(0) = 0$, and $u \in \Re$, is said to be feedback linearizable if there exists a region U in $\Re^n$ containing the origin, a diffeomorphism T: $U \to \Re^n$, and nonlinear feedback

$$u = \alpha(x) + \beta(x)v \tag{16}$$

with $\beta(x) \neq 0$ on U, such that the transformed variables

$$y = T(x) \tag{17}$$

satisfy the system of equations

$$\dot{y} = Ay + bv \tag{18}$$

where

$$A = \begin{bmatrix} 0 & 1 & 0 & & & 0 \\ 0 & 0 & 1 & & & . \\ . & . & . & . & . & . \\ . & . & . & & . & . \\ . & . & . & & & 1 \\ 0 & 0 & . & . & 0 & 0 \end{bmatrix} \quad b = \begin{bmatrix} 0 \\ 0 \\ . \\ . \\ . \\ 1 \end{bmatrix}$$

In the single-link case we see that the appropriate state variables with which to define the system so that it could be linearized by nonlinear feedback on the link position, velocity, acceleration, and jerk. Following the single-input case, then, we can apply same action on the multi-link case and derive a feedback linearizing transformation blockwise as follows,

$$y_1 = T_1(x) = x_1 \tag{19}$$

$$y_2 = T_2(x) = \dot{y}_1 = x_2 \tag{20}$$

$$y_3 = T_3(x) = \dot{y}_2 = \dot{x}_2 = -D(x_1)^{-1}\{h(x_1, x_2) + K(x_1 - x_3)\} \tag{21}$$

$$\begin{aligned} y_4 = T_4(x) = \dot{y}_3 = -\frac{d}{dt}[D(x_1)^{-1}]\{h(x_1, x_2) + K(x_1 - x_3)\} - D(x_1)^{-1}\{\frac{\partial h}{\partial x_1}x_2 \\ +\frac{\partial h}{\partial x_2}[-D(x_1)^{-1}(h(x_1, x_2) + K(x_1 - x_3))] + K(x_2 - x_4)\} = a_4(x_1, x_2, x_3) + D(x_1)^{-1}Kx_4 \end{aligned} \tag{22}$$

where for simplicity we define the function a_4 to be that in the definition of y_4 except the last term, which is $D^{-1}Kx_4$. Note that x_4 appears only in this last term so that a_4 depends only on x_1, x_2, x_3.

As in the single-link case, the above mapping is a global diffeomorphism. Its inverse can be found by

$$x_1 = y_1 \tag{23}$$

$$x_2 = y_2 \tag{24}$$

$$x_3 = y_1 + K^{-1}(D(y_1)y_3 + h(y_1, y_2)) \tag{25}$$

$$x_4 = K^{-1}D(y_1)(y_4 - a_4(y_1, y_2, y_3)) \tag{26}$$

The linearizing control law can now be found from the condition

$$y_4 = v \tag{27}$$

where v is a new control input. Computing $\dot{y}_4$ from (22) and suppressing function arguments for brevity yields

$$v = \frac{\partial a_4}{\partial x_1}x_2 - \frac{\partial a_4}{\partial x_2}D^{-1}(h + K(x_1 - x_3)) + \frac{\partial a_4}{\partial x_3}x_4 + \frac{d}{dt}[D^{-1}]Kx_4 + D^{-1}K(J^{-1}K(x_1 - x_3) + J^{-1}u) \tag{28}$$
$$= a(x) + b(x)u$$

where

$$a(x) =: \frac{\partial a_4}{\partial x_1}x_2 - \frac{\partial a_4}{\partial x_2}D^{-1}(h + K(x_1 - x_3)) + \frac{\partial a_4}{\partial x_3}x_4 + \frac{d}{dt}[D^{-1}]Kx_4 + D^{-1}KJ^{-1}K(x_1 - x_3) \tag{29}$$

$$b(x) = D^{-1}(x)KJ^{-1}u \tag{30}$$

Solving the above expression for u yields

$$u = b(x)^{-1}(v - a(x)) \tag{31}$$

$$=: \alpha(x) + \beta(x)v \tag{32}$$

where $\beta(x) = JK^{-1}D(x)$ and $\alpha(x) = -b(x)^{-1}a(x)$

With the nonlinear change of coordinates (19)-(22) and nonlinear feedback (32) the transformed system has the linear block form

$$\dot{y} = \begin{bmatrix} 0 & I & 0 & 0 \\ 0 & 0 & I & 0 \\ 0 & 0 & 0 & I \\ 0 & 0 & 0 & 0 \end{bmatrix} y + \begin{bmatrix} 0 \\ 0 \\ 0 \\ I \end{bmatrix} v \tag{33}$$
$$=: Ay + bv$$

where $I = n \times n$ identity matrix, $0 = n \times n$ zero matrix, $y^T = (y_1^T, y_2^T, y_3^T, y_4^T) \in \Re^{4n}$, and $v \in \Re^n$. The system (33) represents a set of n decoupled quadruple integrators.

4. Outer loop design based on predictive function control

4.1 why use predictive function control

The technique of feedback linearization is important due to it leads to a control design methodology for nonlinear systems. In the context of control theory, however, one should be highly suspicious of techniques that rely on exact mathematical cancellation of terms, linear or nonlinear, from the equations defining the system.

In this section, we investigate the effect of parameter uncertainty, computational error, model simplification, and etc. We show that the most important property of feedback linearizable systems is not necessarily that the nonlinearities can be exactly cancelled by nonlinear feedback, but rather that, once an appropriate coordinate system is found in which the system can be linearized, the nonlinearities are in the range space of the input. This property is highly significant and is exploited by the predictive function control techniques to guarantee performance in the realistic case that the nonlinearities in the system are not known exactly.

Consider a single-input feedback linearizable system. After the appropriate coordinate transformation, the system can be written in the ideal case as

$$\begin{aligned} \dot{y}_1 &= y_2 \\ &\vdots \\ &\vdots \\ \dot{y}_n &= v = \beta^{-1}(x)[u - \alpha(x)] \end{aligned} \tag{34}$$

provided that u is given by (16) in order to cancel the nonlinear terms $\alpha(x)$ and $\beta(x)$.

In practice such exact cancellation is not achievable and it is more realistic to suppose that the control law u in (16) is of the form

$$u = \hat{\alpha}(x) + \hat{\beta}(x)v \tag{35}$$

where $\hat{\alpha}(x)$, $\hat{\beta}(x)$ represent the computed versions of $\alpha(x)$, $\beta(x)$, respectively. These functions may differ from the true $\alpha(x)$, $\beta(x)$ for several reasons. Because the inner loop control u is implemented digitally, there will be an error due to computational round-off and delay. Also, since the terms $\alpha(x)$, $\beta(x)$ are functions of the system parameters such as masses, and moments of inertia, any uncertainty in knowledge of these parameters will be in reflected in $\hat{\alpha}(x)$, $\hat{\beta}(x)$. In addition, one may choose intentionally to simplify the control u by dropping various terms in the equations in order to facilitate on-line computation. If we now substitute the control law (35) into (34) we obtain

$$\begin{aligned} \dot{y}_1 &= y_2 \\ &\vdots \\ &\vdots \\ \dot{y}_n &= v = \beta^{-1}(x)[\hat{\alpha}(x) + \hat{\beta}(x)v - \alpha(x)] \\ &= v + \eta(y_1, \cdots, y_n, v) \end{aligned} \tag{36}$$

where the uncertainty η is given as

$$\eta(y_1, \cdots, y_n, v) = \left\{\left(\beta^{-1}\hat{\beta} - 1\right)v + \beta^{-1}\left(\hat{\alpha} - \alpha\right)\right\}|_{y=T^{-1}(X)} \tag{37}$$

The system (36) can be written in matrix form as

$$\dot{y} = Ay + b\{v + \eta(y, v)\} \tag{38}$$

where A and b are given by (18). For multi-input case, similar to (33), and if $v \in \Re^m$, and $\eta : \Re^n \times \Re^m \to \Re^m$. Note that the system (38) is still nonlinear whenever $\eta \neq 0$. The practical implication of this is solved by the outer loop predictive function control (PFC).

The system (38) can be represented by the block diagram of Figure 2. The application of the nonlinear inner loop control law results in a system which is "approximately linear". A common approach is to decompose the control input v in (38) into two parts, the first to stabilize the 'nominal linear system' represented by (38) with $\eta = 0$. In this case v can be taken as a linear state feedback control law designed to stabilize the nominal system and/or for tracking a desired trajectory. A second stage control Δv is then designed for robustness, that is, to guarantee the performance of the nominal design in the case that $\eta \neq 0$. Thus the form of the control law is

$$u = \hat{\alpha}(x) + \hat{\beta}(x)v \tag{39}$$

$$v = -Ky + \Delta v \tag{40}$$

$$\Delta v = PFC(y_r) \tag{41}$$

where Ky is a linear feedback designed to place the eigenvalues of A in a desired location, Δv represents an additional feedback loop to maintain the nominal performance despite the presence of the nonlinear term η. y_r is a reference input, which can be chosen as a signal for tracking a desired trajectory.

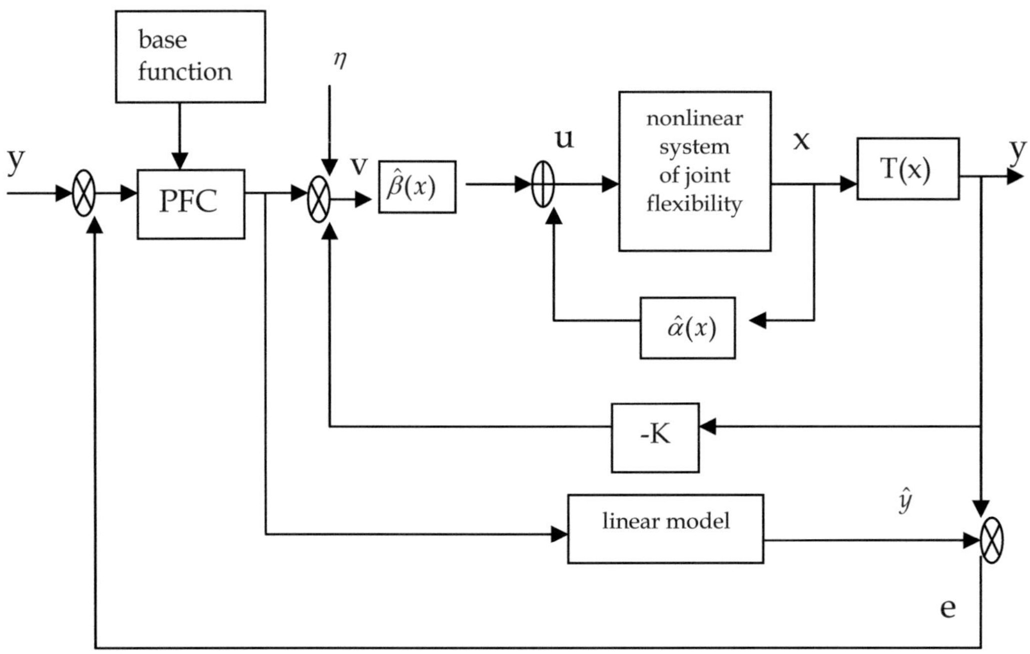

Fig. 2. block diagram for PFC outer loop design

4.2 Predictive function control

All MPC strategies use the same basic approach i.e., prediction of the future plant outputs, and calculation of the manipulated variable for an optimal control. Most MPC strategies are based on the following principles:

Use of an internal model

Its formulation is not restricted to a particular form, and the internal model can be linear, nonlinear, state space form, transfer function form, first principles, black-box etc. In PFC, only independent models where the model output is computed only with the present and past inputs of the process models are used.

Specification of a reference trajectory

Usually an exponential.

Determination of the control law

The control law is derived from the minimization of the error between the predicted output and the reference with the projection of the Manipulated Variable (MV) on a basis of functions.

Although based on these principles the PFC algorithm may be of several levels of complexity depending on the order and form of the internal model, the order of the basis function used to decompose the MV and the reference trajectory used.

4.3 First order PFC

Although it is unrealistic to represent industrial systems by a first order system, as most of them are in a higher order, some well behaved ones may be estimated by a first order. The estimation will not be perfect at each sample time, however, the robustness of the PFC will help to maintain a decent control.

If the system can be modelled by a first order plus pure time delay system, then the following steps in the development of the control law are taken.

Model formulation

In order to implement a basic first order PFC, a typical first order transfer function equation (42) is used.

$$y_M(s) = \frac{K_M}{T_M S + 1} u(s) \tag{42}$$

Note that the time delay is not considered in the internal model formulation and in this case K_M is equal to one. The discrete time formulation of the model zero-order hold equivalent is then obtained in (43).

$$y_M(k) = \alpha y_M(k-1) + K_M(1-\alpha)u(k-1) \tag{43}$$

where $\alpha = \exp(-\frac{T_s}{T_M})$. If the manipulated variable is structured as a step basis function:

$$y_L(k+H) = \alpha^H y_M(k) \tag{44}$$

$$y_F(k+H) = K_M(1-\alpha^H)u(k) \tag{45}$$

Where, y_L and y_F are respectively, the free (autoregressive) and the forced response of y_M .

Reference trajectory formulation

If y_R is the expression of the reference trajectory, then at the coincidence point H:

$$C(k+H)-y_R(k+H)=\lambda^H(C(k)-y_P(k)) \tag{46}$$

thus:

$$y_R(k+H)=C(k)-\lambda^H(C(k)-y_P(k)) \tag{47}$$

Predicted process output

The predicted process output is given by the model response, plus a term given the error between the same model output and the process output:

$$\hat{y}_P(k+H)=y_M(k+H)+(y_P(k)-y_M(k)) \tag{48}$$

where $y_M(k+H)=y_L(k+H)+y_F(k+H)=\alpha^H y_M(k)+K_M(1-\alpha^H)u(k)$.

Computation of the control law

At the coincidence point H:

$$y_R(k+H)=\hat{y}_P(k+H) \tag{49}$$

Combining (44), (45), (47) and (48) yields

$$C(k)-\lambda^H(C(k)-y_P(k))-y_P(k)=y_M(k+H)-y_M(k) \tag{50}$$

Replacing $y_M(k+H)$ by its equivalent in equations (44) and (45) we obtain:

$$C(k)(1-\lambda^H)-y_P(k)(1-\lambda^H)+y_M(k)(1-\alpha^H)=K_M(1-\alpha^H)u(k) \tag{51}$$

Solving for u(k) the final result is the control law given in (52).

$$u(k)=\frac{(C(k)-y_P(k))(1-\lambda^H)}{K_M(1-\alpha^H)}+\frac{y_M(k)}{K_M} \tag{52}$$

4.4 Case of a process with a pure time delay

In the linear case, a process with a pure time delay can be expressed in terms of a delay-free part, plus a delay added at the output, as in Fig. 3.

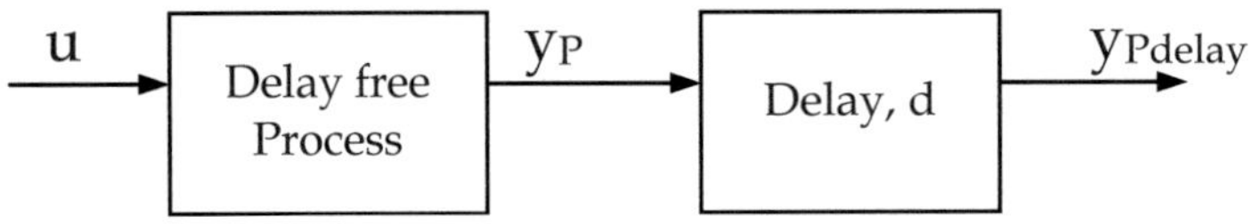

Fig. 3. Process with time delay

The value y_{Pdelay} at time k is measured, but not y_P. In order to take into account the delay in a control law formulation, prior knowledge of the delay value d is needed. y_P can be estimated as:

$$y_P(k) = y_{Pdelay}(k) + y_M(k) - y_M(k-d) \tag{53}$$

4.5 Tuning in PFC

According to the three principles of PFC, tuning is a function of the order of the basis constructing the MV, the reference trajectory, the control horizon and the CLRT value.

The influence of the PFC parameters is given in Table 1, where the influence of various PFC parameters is on precision (Steady State Resp.), transient response and robustness are graded between 0 (indicating minimum influence) and 1 (indicating maximum influence).

	SS Resp.	Transient Resp.	Robustness
Basis function	1	0	0
Reference trajectory	0	1	1/2
Coincidence horizon	0	1/2	1

Table 1. Effect of PFC parameters in tuning

In most cases, an exponential reference trajectory is chosen along with a single coincidence horizon point (H = 1) and a zero order basis function (Richalet, 1993). Considering the known Open Loop Response Time of the system (OLRT), one can choose the CLRT value given by the ratio OLRT/CLRT. This ratio then becomes the major tuning parameter shaping the system output and MV, dictating how much overshoot occurs and ensuring stability, on the condition that the internal model is accurate enough. For slow processes, e.g., heat exchange systems, a ratio of 4 or 5 is found most suitable, and ensures a stable PFC.

5. Simulation

Consider the single link manipulator with flexible joint shown in Figure 1. Choosing q_1 and q_2 as generalized coordinates, the kinetic energy is

$$K = \frac{1}{2} I \dot{q}_1^2 + \frac{1}{2} J \dot{q}_2^2 \tag{54}$$

The potential energy is

$$V = MgL(1 - \cos q_1) + \frac{1}{2} k (q_1 - q_2)^2 \tag{55}$$

The Lagrangian is

$$L = K - V = \frac{1}{2} I \dot{q}_1^2 + \frac{1}{2} J \dot{q}_2^2 - MgL(1 - \cos q_1) - \frac{1}{2} k (q_1 - q_2)^2 \tag{56}$$

Therefore we compute

$$\frac{\partial L}{\partial \dot{q}_1} = I\dot{q}_1 \qquad \frac{\partial L}{\partial \dot{q}_2} = J\dot{q}_2 \tag{57}$$

$$\frac{d}{dt}\frac{\partial L}{\partial \dot{q}_1} = I\ddot{q}_1 \qquad \frac{d}{dt}\frac{\partial L}{\partial \dot{q}_2} = J\ddot{q}_2 \tag{58}$$

$$\frac{\partial L}{\partial q_1} = -MgL\sin(q_1) - k(q_1 - q_2) \qquad \frac{\partial L}{\partial q_2} = k(q_1 - q_2) \tag{59}$$

Therefore the equations of motion, ignoring damping, are given by

$$I\ddot{q}_1 + MgL\sin(q_1) + k(q_1 - q_2) = 0 \tag{60}$$

$$J\ddot{q}_2 - k(q_1 - q_2) = u \tag{61}$$

Note that since the nonlinearity enters into the first equation the control u cannot simply be chosen to cancel it as in the case of the rigid manipulator equations. In other words, there is no obvious analogue of the inverse dynamics control for the system in this form.

In state space we set

$$x_1 = q_1 \qquad x_2 = \dot{q}_1$$

$$x_3 = q_2 \qquad x_4 = \dot{q}_2$$

and write the system (60)- (61) as

$$\begin{aligned}
\dot{x}_1 &= x_2 \\
\dot{x}_2 &= -\frac{MgL}{I}\sin(x_1) - \frac{k}{I}(x_1 - x_3) \\
\dot{x}_3 &= x_4 \\
\dot{x}_4 &= \frac{k}{J}(x_1 - x_3) + \frac{1}{J}u
\end{aligned} \tag{62}$$

The system is thus of the form (15) with

$$f(x) = \begin{bmatrix} x_2 \\ -\frac{MgL}{I}\sin(x_1) - \frac{k}{I}(x_1 - x_3) \\ x_4 \\ \frac{k}{J}(x_1 - x_3) \end{bmatrix} \qquad g(x) = \begin{bmatrix} 0 \\ 0 \\ 0 \\ \frac{1}{J} \end{bmatrix} \tag{63}$$

Therefore n=4 and the necessary and sufficient conditions for feedback linearization of this system are that

$$rank\left\{g, ad_f(g), ad_f^2(g), ad_f^3(g)\right\} = rank\begin{bmatrix} 0 & 0 & 0 & \frac{k}{IJ} \\ 0 & 0 & \frac{k}{IJ} & 0 \\ 0 & \frac{1}{J} & 0 & -\frac{k}{J^2} \\ \frac{1}{J} & 0 & -\frac{k}{J^2} & 0 \end{bmatrix} = 4 \tag{64}$$

which has rank 4 for $k > 0$, $I, J < \infty$. Also, since vector fields $\left\{g, ad_f(g), ad_f^2(g)\right\}$ are constant, they form an involutive set.

$$\left\{g, ad_f(g), ad_f^2(g)\right\} \tag{65}$$

To see this it suffices to note that the Lie Bracket of two constant vector fields is zero. Hence the Lie Bracket of any two members of the set of vector fields in (65) is zero which is trivially a linear combination of the vector fields themselves. It follows that the system (60)- (61) is feedback linearizable. The new coordinates

$$y_i = T_i \quad i = 1, \cdots, 4 \tag{66}$$

are found from the conditions (67)- (68)

$$< dT_1, ad_f^k(g) >= 0 \qquad k = 0, 1, \cdots, n-2 \tag{67}$$

$$< dT_1, ad_f^{n-1}(g) > \neq 0 \tag{68}$$

with n=4, that is

$$< dT_1, g >= 0 \tag{69}$$

$$< dT_1, [f, g] >= 0 \tag{70}$$

$$< dT_1, ad_f^2(g) >= 0 \tag{71}$$

$$< dT_1, ad_f^3(g) > \neq 0 \tag{72}$$

Carrying out the above calculations leads to the system of equations

$$\frac{\partial T_1}{\partial x_2} = 0 \quad \frac{\partial T_1}{\partial x_3} = 0 \quad \frac{\partial T_1}{\partial x_4} = 0 \tag{73}$$

and

$$\frac{\partial T_1}{\partial x_1} \neq 0 \tag{74}$$

From this we see that the function T_1 should be a function of x_1 alone. Therefore, we take the simplest solution

$$y_1 = T_1 = x_1 \tag{75}$$

and compute

$$y_2 = T_2 =< dT_1, f >= x_2 \tag{76}$$

$$y_3 = T_3 =< dT_2, f >= -\frac{MgL}{I}\sin(x_1) - \frac{k}{I}(x_1 - x_3) \tag{77}$$

$$y_4 = T_4 =< dT_3, f >= -\frac{MgL}{I}\cos(x_1) \times x_2 - \frac{k}{I}(x_2 - x_4) \tag{78}$$

The feedback linearizing control input u is found from the condition

$$\begin{aligned} u &= \frac{1}{< dT_4, g >}(v - < dT_4, f >) \\ &= \frac{IJ}{k}(v - a(x)) = \beta(x)v + \alpha(x) \end{aligned} \tag{79}$$

where

$$\alpha(x) := \frac{MgL}{I}\sin(x_1)(x_2^2 + \frac{MgL}{I}\cos(x_1) + \frac{k}{I}) + \frac{k}{I}(x_1 - x_3)(\frac{k}{I} + \frac{k}{J} + \frac{MgL}{I}\cos(x_1)) \tag{80}$$

Therefore in the coordinates $y_1, \cdots, y_4$ with the control law (79) the system becomes

$$\dot{y}_1 = y_2 \quad \dot{y}_2 = y_3$$

$$\dot{y}_3 = y_4 \quad \dot{y}_4 = v$$

or, in matrix form,

$$\dot{y} = Ay + bv \tag{81}$$

where

$$A = \begin{bmatrix} 0 & 1 & 0 & 0 \\ 0 & 0 & 1 & 0 \\ 0 & 0 & 0 & 1 \\ 0 & 0 & 0 & 0 \end{bmatrix} \qquad b = \begin{bmatrix} 0 \\ 0 \\ 0 \\ 1 \end{bmatrix}$$

The transformed variables $y_1, \cdots, y_4$ are themselves physically meaningful. We see that

$$y_1 = x_1 = \text{link position}$$

$$y_2 = x_2 = \text{link velocity}$$

$$y_3 = \dot{y}_2 = \text{link acceleration}$$

$$y_4 = \dot{y}_3 = \text{link jerk}$$

Since the motion trajectory of the link is typically specified in terms of these quantities they are natural variables to use for feedback.

For given a linear system in state space form, such as (81), a state feedback control law is an input v of the form

$$v = -k^T y + r = -\sum_{i=1}^{4} k_i y_i + r \tag{82}$$

where k_i are constants and r is a reference input. If we substitute the control law (82) into (81), we obtain

$$\dot{y} = (A - bk^T)y + br \tag{83}$$

Thus we see that the linear feedback control has the effect of changing the poles of the system from those determined by A to those determined by $A - bk^T$

When the parameters are chosen $k_1 = 62.5$, $k_2 = 213.6$, $k_3 = 204.2$ $k_4 = 54$, we can get step responses in Figure 4. where k1, k2, k3 and k4 are linear feedback coefficients to place the eigenvalues of A in a desired location.

$$\begin{bmatrix} \dot{y}_1 \\ \dot{y}_2 \\ \dot{y}_3 \\ \dot{y}_4 \end{bmatrix} = \begin{bmatrix} 0 & 1 & 0 & 0 \\ 0 & 0 & 1 & 0 \\ 0 & 0 & 0 & 1 \\ -62.5 & -213.8 & -204.2 & -54 \end{bmatrix} \begin{bmatrix} y_1 \\ y_2 \\ y_3 \\ y_4 \end{bmatrix} + \begin{bmatrix} 0 \\ 0 \\ 0 \\ 1 \end{bmatrix} r \tag{84}$$

$$y = \begin{bmatrix} 1 & 0 & 0 & 0 \end{bmatrix} \begin{bmatrix} y_1 \\ y_2 \\ y_3 \\ y_4 \end{bmatrix} \tag{85}$$

The internal model parameter: $K_M = 0.016$, $T_M = 3$, $d = 8$, and the coincidence point H=10.

Response time of reference trajectory is 0.01, and sample time is 0.01.

For the uncertainty η, the system (38) can be written in matrix form as

$$\dot{y} = (A - bk^T)y + b\{r + \eta(y, v)\}$$

then use predictive function control strategy to reduce or overcome uncertainty of nonlinear feedback error $\eta(y, v)$,and simulation result is shown in Figure 5 for $\eta(y, v) = 10\% y_r$.

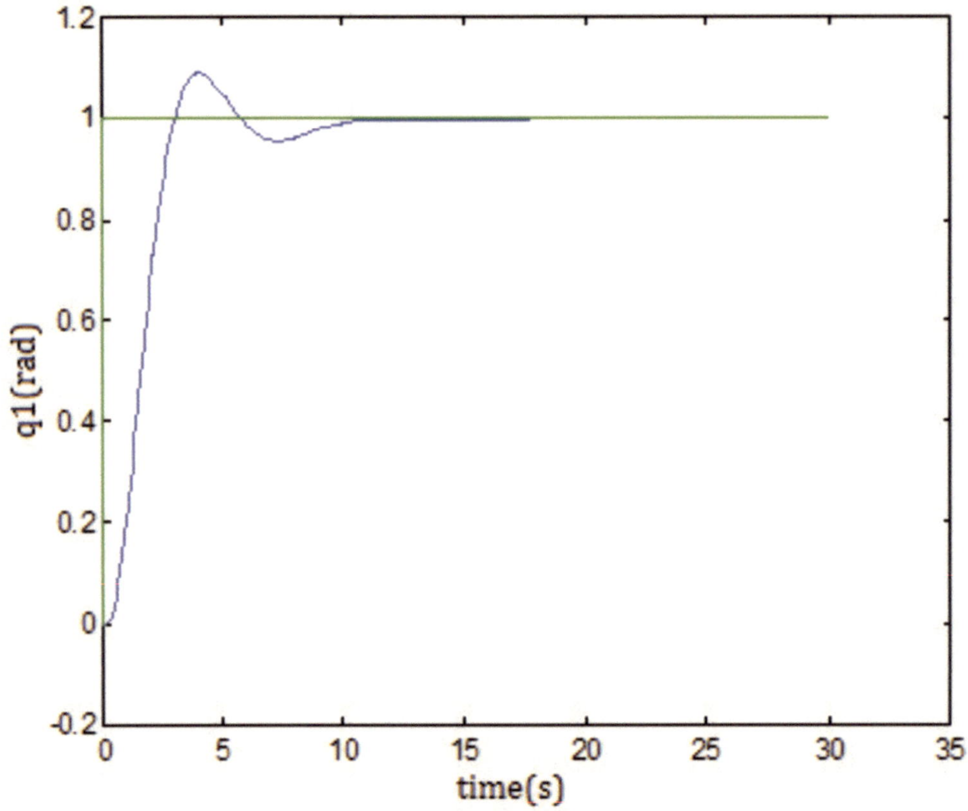

Fig. 4. link position output

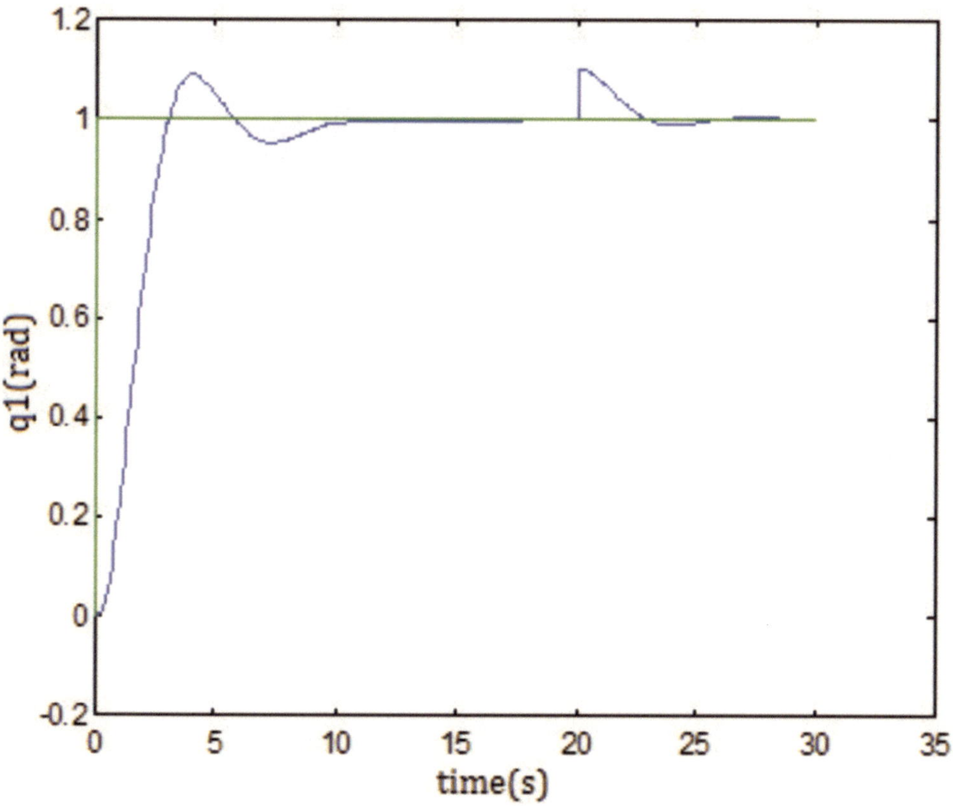

Fig. 5. link position output with uncertainty rejection

6. Conclusion

A new three stage design method is presented for the single link manipulator with flexible joint. The first is feedback linearization; the second is to use pole placement to satisfy performance, and the third is to develop predictive function control to compensate uncertainty. Finally, for the same uncertainty, robustness is better than traditional method.

7. References

Khorasani, K. (1990). Nonlinear feedback control of flexible joint manipulators: a single link case study. *IEEE Transactions on Automatic Control,* Vol. 35, No. 10, pp. 1145-1149, ISSN 0018-9286

Richalet, J.; Rault, A.; Testud, J. L.; Papon, J.(1978). Model predictive heuristic control: Applications to industrial processes. *Automatica,* Vol. 14, No. 5, pp. 413-428, ISSN 0005-1098

Spong, M.W.; Khorasani, K. & Kokotovic P.V. (1987). An integral manifold approach to the feedback control of flexible joint robots. *IEEE Journal of Robotics and Automation,* Vol. RA-3, No. 4, pp. 291-300, ISSN 0882-4967

Spong, M.W.(1987). Modeling and control of elastic joint robots. *Journal of Dyn. Sys. Meas. and Cont.,* Vol. 109, pp.310-319, ISSN 0022-0434

Spong, M.W. (1989). On the force control problem for flexible joint manipulator. *IEEE Transactions on Automatic Control,* Vol. 34, No. 1, pp. 107-111, ISSN 0018-9286

E. F. Camacho, Carlos Bordons(2004). *Model predictive control.* Springer-Verlag London Berlin Limited. ISBN 1-85233-694-3

9

On Optimization Techniques for a Class of Hybrid Mechanical Systems

Vadim Azhmyakov and Arturo Enrique Gil García
Department of Control Automation, CINVESTAV, A.P. 14-740, Mexico D.F., Mexico

1. Introduction

Several classes of general hybrid and switched dynamic systems have been extensively studied, both in theory and practice [3,4,7,11,14,17,19,26,27,30]. In particular, driven by engineering requirements, there has been increasing interest in optimal design for hybrid control systems [3,4,7,8,13,17,23,26,27]. In this paper, we investigate some specific types of hybrid systems, namely hybrid systems of mechanical nature, and study the corresponding hybrid OCPs. The class of dynamic models to be discussed in this work concerns hybrid systems where discrete transitions are being triggered by the continuous dynamics. The control objective (control design) is to minimize a cost functional, where the control parameters are the conventional control inputs.

Recently, there has been considerable effort to develop theoretical and computational frameworks for complex control problems. Of particular importance is the ability to operate such systems in an optimal manner. In many real-world applications a controlled mechanical system presents the main modeling framework and can be specified as a strongly nonlinear dynamic system of high order [9,10,22]. Moreover, the majority of applied OCPs governed by sophisticated real-world mechanical systems are optimization problems of the hybrid nature. The most real-world mechanical control problems are becoming too complex to allow analytical solution. Thus, computational algorithms are inevitable in solving these problems. There is a number of results scattered in the literature on numerical methods for optimal control problems. One can find a fairly complete review in [3,4,8,24,25,29]. The main idea of our investigations is to use the variational structure of the solution to the specific two point boundary-value problem for the controllable hybrid-type mechanical systems in the form of Euler-Lagrange or Hamilton equation and to propose a new computational algorithm for the associated OCP. We consider an OCP in mechanics in a general setting and reduce the initial problem to a constrained multiobjective programming. This auxiliary optimization approach provides a basis for a possible numerical treatment of the original problem.

The outline of our paper is as follows. Section 2 contains some necessary basic facts related to the conventional and hybrid mechanical models. In Section 3 we formulate and study our main optimization problem for hybrid mechanical systems. Section 4 deals with the variational analysis of the OCP under consideration. We also briefly discuss the computational aspect of the proposed approach. In Section 5 we study a numerical example that constitutes an implementable hybrid mechanical system. Section 6 summarizes our contribution.

2. Preliminaries and some basic facts

Let us consider the following variational problem for a hybrid mechanical system that is characterized by a family of Lagrange functions $\{\tilde{L}_{p_i}\}$, $p_i \in \mathcal{P}$

$$\begin{aligned} &\text{minimize} \int_0^1 \sum_{i=1}^{r} \beta_{[t_{i-1},t_i)}(t)\tilde{L}_{p_i}(t, q(t), \dot{q}(t))dt \\ &\text{subject to } q(0) = c_0,\ q(1) = c_1, \end{aligned} \tag{1}$$

where $\mathcal{P}$ is a finite set of indices (locations) and $q(\cdot)$ ($q(t) \in \mathbb{R}^n$) is a continuously differentiable function. Here $\beta_{[t_{i-1},t_i)}(\cdot)$ are characteristic functions of the time intervals $[t_{i-1}, t_i)$, $i = 1, ..., r$ associated with locations. Note that a full time interval $[0, 1]$ is assumed to be separated into disjunct sub-intervals of the above type for a sequence of switching times:

$$\tau := \{t_0 = 0, t_1, ..., t_r = 1\}.$$

We refer to [3,4,7,8,13,17,23,26,27] for some concrete examples of hybrid systems with the above dynamic structure. Consider a class of hybrid mechanical systems that can be represented by n generalized configuration coordinates $q_1, ..., q_n$. The components $\dot{q}_\lambda(t), \lambda = 1, ..., n$ of $\dot{q}(t)$ are the so-called generalized velocities. Moreover, we assume that $\tilde{L}_{p_i}(t, \cdot, \cdot)$ are twice continuously differentiable convex functions. It is well known that the formal necessary optimality conditions for the given variational problem (1) describe the dynamics of the mechanical system under consideration. This description can be given for every particular location and finally, for the complete hybrid system. In this contribution, we study the hybrid dynamic models that free from the possible external influences (uncertainties) or forces. The optimality conditions for mentioned above can be rewritten in the form of the second-order Euler-Lagrange equations (see [1])

$$\begin{aligned} &\frac{d}{dt}\frac{\partial \tilde{L}_{p_i}(t, q, \dot{q})}{\partial \dot{q}_\lambda} - \frac{\partial \tilde{L}_{p_i}(t, q, \dot{q})}{\partial q_\lambda} = 0,\ \lambda = 1, ..., n\,, \\ &q(0) = c_0,\ q(1) = c_1, \end{aligned} \tag{2}$$

for all $p_i \in \mathcal{P}$. The celebrated Hamilton Principle (see e.g., [1]) gives an equivalent variational characterization of the solution to the two-point boundary-value problem (2).

For the controllable hybrid mechanical systems with the parametrized (control inputs) Lagrangians $L_{p_i}(t, q, \dot{q}, u)$, $p_i \in \mathcal{P}$ we also can introduce the corresponding equations of motion

$$\begin{aligned} &\frac{d}{dt}\frac{\partial L_{p_i}(t, q, \dot{q}, u)}{\partial \dot{q}_\lambda} - \frac{\partial L_{p_i}(t, q, \dot{q}, u)}{\partial q_\lambda} = 0, \\ &q(0) = c_0,\ q(1) = c_1, \end{aligned} \tag{3}$$

where $u(\cdot) \in \mathcal{U}$ is a control function from the set of admissible controls $\mathcal{U}$. Let

$$U := \{u \in \mathbb{R}^m \ : \ b_{1,\nu} \leq u_\nu \leq b_{2,\nu},\ \nu = 1, ..., m\},$$
$$\mathcal{U} := \{v(\cdot) \in \mathbb{L}^2_m([0,1]) \ : \ v(t) \in U \text{ a.e. on } [0,1]\},$$

where $b_{1,\nu}, b_{2,\nu}, \nu = 1, ..., m$ are constants. The introduced set $\mathcal{U}$ provides a standard example of an admissible control set. In this specific case we deal with the following set of admissible controls $\mathcal{U} \bigcap \mathbb{C}^1_m(0,1)$. Note that L_{p_i} depends directly on the control function $u(\cdot)$. Let us assume that functions $L_{p_i}(t,\cdot,\cdot,u)$ are twice continuously differentiable functions and every $L_{p_i}(t,q,\dot{q},\cdot)$ is a continuously differentiable function. For a fixed admissible control $u(\cdot)$ we obtain for all $p_i \in \mathcal{P}$ the above hybrid mechanical system with $\tilde{L}_{p_i}(t,q,\dot{q}) \equiv L_{p_i}(t,q,\dot{q},u(t))$. It is also assumed that $L_{p_i}(t,q,\cdot,u)$ are strongly convex functions, i.e., for any

$$(t,q,\dot{q},u) \in \mathbb{R} \times \mathbb{R}^n \times \mathbb{R}^n \times \mathbb{R}^m, \ \ \xi \in \mathbb{R}^n$$

the following inequality

$$\sum_{\lambda,\theta=1}^{n} \frac{\partial^2 L_{p_i}(t,q,\dot{q},u)}{\partial \dot{q}_\lambda \partial \dot{q}_\theta} \xi_\lambda \xi_\theta \geq \alpha \sum_{\lambda=1}^{n} \xi^2_\lambda,\ \alpha > 0$$

holds for all $p_i \in \mathcal{P}$. This natural convexity condition is a direct consequence of the classical representation for the kinetic energy of a conventional mechanical system. Under the above-mentioned assumptions, the two-point boundary-value problem (3) has a solution for every admissible control $u(\cdot) \in \mathcal{U}$ [18]. We assume that (3) has a unique solution for every $u(\cdot) \in \mathcal{U}$. For an admissible control $u(\cdot) \in \mathcal{U}$ the solution to the boundary-value problem (3) is denoted by $q^u(\cdot)$. We call (3) the hybrid *Euler-Lagrange control system*. Let us now give a simple example of the above mechanical model.

Example 1. *We consider a variable linear mass-spring system attached to a moving frame that is a generalization of the corresponding system from [22]. The considered control $u(\cdot) \in \mathcal{U} \bigcap \mathbb{C}^1_1(0,1)$ is the velocity of the frame. By ω_{p_i} we denote the variable masses of the system. The kinetic energy*

$$K = 0.5\omega_{p_i}(\dot{q} + u)^2$$

depends on the control input $u(\cdot)$. Therefore, we have

$$L_{p_i}(q,\dot{q},u) = 0.5(\omega_{p_i}(\dot{q} + u)^2 - \kappa q^2),\ \kappa \in \mathbb{R}_+$$

and

$$\frac{d}{dt}\frac{\partial L_{p_i}(t,q,\dot{q},u)}{\partial \dot{q}} - \frac{\partial L_{p_i}(t,q,\dot{q},u)}{\partial q} = \omega_{p_i}(\ddot{q} + \dot{u}) + \kappa q = 0.$$

By κ we denote here the elasticity coefficient of the spring system.

Note that some important controlled mechanical systems have a Lagrangian function of the following form (see e.g., [22])

$$L_{p_i}(t,q,\dot{q},u) = L^0_{p_i}(t,q,\dot{q}) + \sum_{\nu=1}^{m} q_\nu u_\nu.$$

In this special case we easily obtain

$$\frac{d}{dt}\frac{\partial L^0_{p_i}(t,q,\dot q)}{\partial \dot q_\lambda} - \frac{\partial L^0_{p_i}(t,q,\dot q)}{\partial q_\lambda} = \begin{cases} u_\lambda & \lambda = 1,...,m, \\ 0 & \lambda = m+1,...,n. \end{cases}$$

Note that the control function $u(\cdot)$ is interpreted here as an external force.

Let us now consider the Hamiltonian reformulation for the controllable Euler-Lagrange system (3). For every location p_i from $\mathcal{P}$ we introduce the generalized momenta

$$s_\lambda := L_{p_i}(t,q,\dot q,u)/\partial \dot q_\lambda$$

and define the Hamiltonian function $H_{p_i}(t,q,s,u)$ as a Legendre transform applied to every $L_{p_i}(t,q,\dot q,u)$, i.e.

$$H_{p_i}(t,q,s,u) := \sum_{\lambda=1}^{n} s_\lambda \dot q_\lambda - L_{p_i}(t,q,\dot q,u).$$

In the case of hyperregular Lagrangians $L_{p_i}(t,q,\dot q,u)$ (see e.g., [1]) the Legendre transform, namely, the mapping

$$\mathcal{L}_{p_i} : (t,q,\dot q,u) \to (t,q,s,u),$$

is a diffeomorphism for every $p_i \in \mathcal{P}$,. Using the introduced Hamiltonian $H(t,q,s,u)$, we can rewrite system (3) in the following Hamilton-type form

$$\begin{aligned} \dot q_\lambda(t) &= \frac{\partial H_{p_i}(t,q,s,u)}{\partial s_\lambda},\ q(0)=c_0,\ q(1)=c_1, \\ \dot s_\lambda(t) &= -\frac{H_{p_i}(t,q,s,u)}{\partial q_\lambda},\ \lambda = 1,...,n\,. \end{aligned} \tag{4}$$

Under the above-mentioned assumptions, the boundary-value problem (4) has a solution for every $u(\cdot) \in \mathcal{U}$. We will call (4) a *Hamilton control system*. The main advantage of (4) in comparison with (3) is that (4) immediately constitutes a control system in standard state space form with state variables (q,s) (in physics usually called the *phase variables*). Consider the system of Example 1 with

$$H_{p_i}(q,s,u) = \frac{1}{2}\omega_{p_i}(\dot q^2 - u^2) + \frac{1}{2}\kappa q^2 - su.$$

The Hamilton equations in this case are given as follows

$$\begin{aligned} \dot q &= \frac{\partial H_{p_i}(q,s,u)}{\partial s} = \frac{1}{\omega_{p_i}}s - u, \\ \dot s &= -\frac{\partial H_{p_i}(q,s,u)}{\partial q} = -\kappa q. \end{aligned}$$

Clearly, for

$$L_{p_i}(t,q,\dot q,u) = L^0_{p_i}(t,q,\dot q) + \sum_{\nu=1}^{m} q_\nu, u_\nu$$

we obtain the associated Hamilton functions in the form

$$H_{p_i}(t,q,s,u) = H^0_{p_i}(t,q,s) - \sum_{\nu=1}^{m} q_\nu u_\nu,$$

where $H^0_{p_i}(t,q,s)$ is the Legendre transform of $L^0_{p_i}(t,q,\dot{q})$.

3. Optimization of control processes in hybrid mechanical systems

Let us formally introduce the class of OCPs investigated in this paper:

$$\begin{aligned} &\text{minimize } J := \int_0^1 \sum_{i=1}^{r} \beta_{[t_{i-1},t_i)}(t) f^0_{p_i}(q^u(t),u(t))dt \\ &\text{subject to } u(t) \in U\, t \in [0,1],\ t_i \in \tau,\ i = 1,...,r, \end{aligned} \tag{5}$$

where $f^0_{p_i} : [0,1] \times \mathbb{R}^n \times \mathbb{R}^m \to \mathbb{R}$ be continuous and convex on $\mathbb{R}^n \times \mathbb{R}^m$ objective functions. We have assumed that the boundary-value problems (3) have a unique solution $q^u(\cdot)$ and that the optimization problem (5) also has a solution. Let $(q^{opt}(\cdot), u^{opt}(\cdot))$ be an optimal solution of (5). Note that we can also use the associated Hamiltonian-type representation of the initial OCP (5). We mainly focus our attention on the application of direct numerical algorithms to the hybrid optimization problem (5). A great amount of works is devoted to the direct or indirect numerical methods for conventional and hybrid OC problems (see e.g., [8,24,25,29] and references therein). Evidently, an OC problem involving ordinary differential equations can be formulated in various ways as an optimization problem in a suitable function space and solved by some standard numerical algorithms (e.g., by applying a first-order methods [3,24]).

Example 2. *Using the* Euler-Lagrange *control system from* Example 1, *we now examine the following OCP*

$$\begin{aligned} &\text{minimize } J := -\int_0^1 \sum_{i=1}^{r} \beta_{[t_{i-1},t_i)}(t) k_{p_i}(u(t) + q(t))dt \\ &\text{subject to } \ddot{q}(t) + \frac{\kappa}{\omega_{p_i}} q(t) = -\dot{u}(t),\ i = 1,...,r, \\ &q(0) = 0,\ q(1) = 1, \\ &u(\cdot) \in C^1_1(0,1),\ 0 \le u(t) \le 1\ \forall t \in [0,1], \end{aligned}$$

where k_{p_i} are given (variable) coefficients. Let $\omega_{p_i} \ge 4\kappa/\pi^2$ for every $p_i \in \mathcal{P}$. The solution $q^u(\cdot)$ of the associated boundary-value problem can be written as follows

$$q^u(t) = C^u_i \sin\left(t\sqrt{\kappa/\omega_{p_i}}\right) - \int_0^t \sqrt{\kappa/\omega_{p_i}} \sin\left(\sqrt{\kappa/\omega_{p_i}}(t-l)\right)\dot{u}(l)dl,$$

where $t \in [t_{i-1},t_i)$, $i = 1,...,r$ and

$$C^u_i = \frac{1}{\sin\sqrt{\kappa/\omega_{p_i}}}\left[1 + \int_0^1 \sqrt{\kappa/\omega_{p_i}} \sin\left(\sqrt{\kappa/\omega}(t-l)\right)\dot{u}(l)dl\right]$$

is a constant in every location. Consequently, we have

$$J = -\int_0^1 \sum_{i=1}^r \beta_{[t_{i-1},t_i)}(t) k_{p_i}[u(t) + q^u(t)]dt =$$

$$-\int_0^1 \sum_{i=1}^r \beta_{[t_{i-1},t_i)}(t) k_{p_i}[u(t) + C_i^u \sin(t\sqrt{\kappa/\omega_{p_i}})$$

$$-\int_0^t \sqrt{\kappa/\omega_{p_i}} \sin(\sqrt{\kappa/\omega_{p_i}}(t-l))\dot{u}(l)dl]dt.$$

Let now $k_{p_i} = 1$ *for all* $p_i \in \mathcal{P}$. *Using the Hybrid Maximum Principle (see [4]), we conclude that the admissible control* $u^{opt}(t) \equiv 0.5$ *is an optimal solution of the given OCP. Note that this result is also consistent with the* Bauer *Maximum Principle (see e.g., [2]). For* $u^{opt}(\cdot)$ *we can compute the corresponding optimal trajectory given as follows*

$$q^{opt}(t) = \frac{\sin(t\sqrt{\kappa/\omega_{p_i}})}{\sin\sqrt{\kappa/\omega_{p_i}}}, \quad t \in [t_{i-1}, t_i), \quad i = 1, ..., r.$$

Evidently, we have $\sqrt{\kappa/\omega_{p_i}} \leq \pi/2$ *for every location* p_i. *Moreover,* $q^{opt}(t) \leq 3$ *under the condition* $q^{opt}(\cdot) \in \mathbb{C}_1^1(0,1)$.

Finally, note that a wide family of classical impulsive control systems (see e.g., [12]) can be described by the conventional controllable Euler-Lagrange or Hamilton equations (see [5]). Moreover, we refer to [6] for impulsive hybrid control systems and associated OCPs. Thus the impulsive hybrid systems of mechanical nature can also be incorporated into the modeling framework presented in this section.

4. The variational approach to hybrid OCPs of mechanical nature

An effective numerical procedure, as a rule, uses the specific structure of the problem under consideration. Our aim is to study the variational structure of the main OCP (5). Let

$$\Gamma_i := \{\gamma(\cdot) \in \mathbb{C}_n^1([t_{i-1}, t_i]) \; : \; \gamma(t_{i-1}) = c_{i-1}, \; \gamma(t_i) = c_i\},$$

where $i = 1, ..., r$.. The vectors c_i, where $i = 1, ..., r$ are defined by the corresponding switching mechanism of a concrete hybrid system. We refer to [3,4,26] for some possible switching rules determined for various classes of hybrid control systems. We now present an immediate consequence of the classical Hamilton Principle from analytical mechanics.

Theorem 1. *Let all* Lagrangians $L_{p_i}(t, q, \dot{q}, u)$ *be a strongly convex function with respect to the generalized variables* $\dot{q}_i$, $i = 1, ..., n$. *Assume that every boundary-value problem from* (3) *has a unique solution for every* $u(\cdot) \in \mathcal{U} \bigcap \mathbb{C}_m^1(0,1)$. *A function* $q^u(\cdot)$, *where* $u(\cdot) \in \mathcal{U} \bigcap \mathbb{C}_m^1(0,1)$, *is a solution of the sequence of boundary-value problems* (3) *if and only if a restriction of this function on every time interval* $[t_{i-1} t_i)$, $i = 1, ..., r$ *can be found as follows*

$$q_i^u(\cdot) = \text{argmin}_{q(\cdot) \in \Gamma_i} \int_{t_{i-1}}^{t_i} L_{p_i}(t, q(t), \dot{q}(t), u(t))dt.$$

For an admissible control function $u(\cdot)$ from $\mathcal{U}$ we now introduce the following two functionals

$$T_{p_i}(q(\cdot),z(\cdot)) := \int_{t_{i-1}}^{t_i} [L_{p_i}(t,q(t),\dot{q}(t),u(t)) - L_{p_i}(t,z(t),\dot{z}(t),u(t))]dt,$$

$$V_{p_i}(q(\cdot)) := \max_{z(\cdot)\in\Gamma_i} \int_{t_{i-1}}^{t_i} [L_{p_i}(t,q(t),\dot{q}(t),u(t)) - L_{p_i}(t,z(t),\dot{z}(t),u(t))]dt,$$

for all indexes $p_i \in \mathcal{P}$. Generally, we define $q^u(\cdot)$ as an element of the Sobolev space $\mathbb{W}_n^{1,\infty}(0,1)$, i.e., the space of absolutely continuous functions with essentially bounded derivatives. Let us give a variational interpretation of the admissible solutions $q^u(\cdot)$ to a sequence of problems (3).

Theorem 2. *Let all* Lagrangians $L_{p_i}(t,q,\dot{q},u)$ *be strongly convex functions with respect to the variables* $\dot{q}_i$, $i=1,...,n$. *Assume that every boundary-value problem from* (3) *has a unique solution for every* $u(\cdot) \in \mathcal{U}\bigcap \mathbb{C}_m^1(0,1)$. *An absolutely continuous function* $q^u(\cdot)$, *where* $u(\cdot) \in \mathcal{U}\bigcap \mathbb{C}_m^1(0,1)$, *is a solution of the sequence of problems* (3) if and only if a restriction of this function on $[t_{i-1}t_i)$, $i=1,...,r$ can be found as follows

$$q_i^u(\cdot) = \text{argmin}_{q(\cdot)\in\mathbb{W}_n^{1,\infty}(t_{i-1},t_i)} V_{p_i}(q(\cdot)) \tag{6}$$

Proof. Let $q^u(\cdot) \in \mathbb{W}_n^{1,\infty}(t_{i-1},t_i)$ be a unique solution of a partial problem (3) on the corresponding time interval, where $u(\cdot) \in \mathcal{U}\bigcap \mathbb{C}_m^1(0,1)$. Using the Hamilton Principle in every location $p_i\ in \mathcal{P}$, we obtain the following relations

$$\min_{q(\cdot)\in\mathbb{W}_n^{1,\infty}(t_{i-1},t_i)} V_{p_i}(q(\cdot)) = \min_{q(\cdot)\in\mathbb{W}_n^{1,\infty}(t_{i-1},t_i)} \max_{z(\cdot)\in\Gamma_i} \int_{t_{i-1}}^{t_i} [L_{p_i}(t,q(t),\dot{q}(t),u(t)) -$$

$$\int_{t_{i-1}}^{t_i} L_{p_i}(t,z(t),\dot{z}(t),u(t))]dt = \min_{q(\cdot)\in\mathbb{W}_n^{1,\infty}(t_{i-1},t_i)} \int_{t_{i-1}}^{t_i} L_{p_i}(t,q(t),\dot{q}(t),u(t))dt -$$

$$\min_{z(\cdot)\in\Gamma_i} \int_{t_{i-1}}^{t_i} L_{p_i}(t,z(t),\dot{z}(t),u(t))dt = \int_{t_{i-1}}^{t_i} L_{p_i}(t,q^u(t),\dot{q}^u(t),u(t))dt -$$

$$\int_{t_{i-1}}^{t_i} L_{p_i}(t,q^u(t),\dot{q}^u(t),u(t))dt = V_{p_i}(q^u(\cdot)) = 0.$$

If the condition (6) is satisfied, then $q^u(\cdot)$ is a solution of the sequence of the boundary-value problem (3). This completes the proof. □

Theorem 1 and Theorem 2 make it possible to express the initial OCP (5) as a multiobjective optimization problem over the set of admissible controls and generalized coordinates

$$\begin{aligned} &\text{minimize } J(q(\cdot),u(\cdot)) \text{ and } P(q(\cdot)) \\ &\text{subject to} \\ &(q(\cdot),u(\cdot)) \in (\bigcup_{i=1,...,r} \Gamma_i) \times (\mathcal{U}\bigcap \mathbb{C}_m^1(0,1)), \end{aligned} \tag{7}$$

or

$$\text{minimize } J(q(\cdot),u(\cdot)) \text{ and } V(q(\cdot))$$
$$\text{subject to} \tag{8}$$
$$(q(\cdot),u(\cdot)) \in (\bigcup_{i=1,\ldots,r} \Gamma_i) \times (\mathcal{U} \bigcap \mathbb{C}^1_m(0,1)),$$

where

$$P(q(\cdot)) := \int_0^1 \sum_{i=1}^{r} \beta_{[t_{i-1},t_i)}(t) L_{p_i}(t,q(t),\dot{q}(t),u^{opt}(t))dt$$

and

$$V(q(\cdot)) := \beta_{[t_{i-1},t_i)}(t) V_{p_i}(q(\cdot)).$$

The auxiliary minimizing problems (7) and (8) are multiobjective optimization problems (see e.g., [16,28]). Note that the set

$$\Gamma \times (\mathcal{U} \bigcap \mathbb{C}^1_m(0,1)$$

is a convex set. Since $f_0(t,\cdot,\cdot)$, $t \in [0,1]$ is a convex function, $J(q(\cdot),u(\cdot))$ is also convex. If $P(\cdot)$ (or $V(\cdot)$) is a convex functional, then we deal with a convex multiobjective minimization problem (7) (or (8)).

The variational representation of the solution of the two-point boundary-value problem (3) eliminates the differential equations from the consideration. The minimization problems (7) and (8) provide a basis for numerical algorithms to the initial OCP (5). The auxiliary optimization problem (7) has two objective functionals. For (7) we now introduce the Lagrange function [28]

$$\Lambda(t,q(\cdot),u(\cdot),\mu,\mu_3) := \mu_1 J(q(\cdot),u(\cdot)) + \mu_2 P(q(\cdot)) +$$
$$\mu_3 |\mu| \text{dist}_{(\bigcup_{i=1,\ldots,r}\Gamma_i)\times(\mathcal{U}\bigcap \mathbb{C}^1_m(0,1))}\{(q(\cdot),u(\cdot))\},$$

where $\text{dist}_{(\bigcup_{i=1,\ldots,r}\Gamma_i)\times(\mathcal{U}\bigcap \mathbb{C}^1_m(0,1))}\{\cdot\}$ denotes the distance function

$$\text{dist}_{(\Gamma_i)\times(\mathcal{U}\bigcap \mathbb{C}^1_m(0,1))}\{(q(\cdot),u(\cdot))\} := \inf\{||(q(\cdot),u(\cdot)) -$$
$$- \varrho||_{\mathbb{C}^1_n(0,1)\times\mathbb{C}^1_m(0,1)}, \ \varrho \in (\bigcup_{i=1,\ldots,r} \Gamma_i) \times (\mathcal{U} \bigcap \mathbb{C}^1_m(0,1))\}.$$

We also used the following notation

$$\mu := (\mu_1,\mu_2)^T \in \mathbb{R}^2_+.$$

Note that the above distance function is associated with the Cartesian product

$$(\bigcup_{i=1,\ldots,r} \Gamma_i) \times (\mathcal{U} \bigcap \mathbb{C}^1_m(0,1)).$$

Recall that a feasible point $(q^*(\cdot), u^*(\cdot))$ is called *weak Pareto optimal* for the multiobjective problem (8) if there is no feasible point $(q(\cdot), u(\cdot))$ for which

$$J(q(\cdot), u(\cdot)) < J(q^*(\cdot), u^*(\cdot)) \text{ and } P(q(\cdot)) < P(q^*(\cdot)).$$

A necessary condition for $(q^*(\cdot), u^*(\cdot))$ to be a weak Pareto optimal solution to (8) in the sense of Karush-Kuhn-Tucker (KKT) condition is that for every $\mu_3 \in \mathbb{R}$ sufficiently large there exist $\mu^* \in \mathbb{R}^2_+$ such that

$$0 \in \partial_{(q(\cdot),u(\cdot))} \Lambda(t, q^*(\cdot), u^*(\cdot), \mu^*, \mu_3). \tag{9}$$

By $\partial_{(q(\cdot),u(\cdot))}$ we denote here the *generalized gradient* of the Lagrange function Λ. We refer to [28] for further theoretical details. If $P(\cdot)$ is a convex functional, then the necessary condition (9) is also sufficient for $(q^*(\cdot), u^*(\cdot))$ to be a weak Pareto optimal solution to (8). Let $\aleph$ be a set of all weak Pareto optimal solutions $(q^*(\cdot), u^*(\cdot))$ for problem (7). Since $(q^{opt}(\cdot)u^{opt}(\cdot)) \in \aleph$, the above conditions (9) are satisfied also for this optimal pair $(q^{opt}(\cdot)u^{opt}(\cdot))$.

It is a challenging issue to develop necessary optimality conditions for the proper Pareto optimal (efficient) solutions. A number of theoretical papers concerning multiobjective optimization are related to this type of Pareto solutions. One can find a fairly complete review in [20]. Note that the formulation of the necessary optimality conditions (9) involves the Clarke generalized gradient of the Lagrange function. On the other hand, there are more effective necessary conditions for optimality based on the concept of the Mordukhovich limiting subdifferentials [20]. The use of the above-mentioned Clarke approach is motivated here by the availability of the corresponding powerful software packages.

When solving constrained optimization based on some necessary conditions for optimality one is often faced with a technical difficulty, namely, with the irregularity of the Lagrange multiplier associated with the objective functional [15,20]. Various supplementary conditions (constraint qualifications) have been proposed under which it is possible to assert that the Lagrange multiplier rule holds in "normal" form, i.e., that the first Lagrange multiplier is nonequal to zero. In this case we call the corresponding minimization problem *regular*. Examples of the constraint qualifications are the well known Slater (regularity) condition for classic convex programming and the Mangasarian-Fromovitz regularity conditions for general nonlinear optimization problems. We refer to [15,20] for details. In the case of a conventional multiobjective optimization problem the corresponding regularity conditions can be given in the form of so called KKT constraint qualification (see [28] for details). In the following, we assume that problems (7) and (8) are regular.

Consider now the numerical aspects of the solution procedure associated with (7) and recall that discrete approximation techniques have been recognized as a powerful tool for solving optimal control problems [3,25,29]. Our aim is to use a discrete approximation of (7) and to obtain a finite-dimensional auxiliary optimization problem. Let N be a sufficiently large positive integer number and

$$\mathcal{G}_i^N := \{t_0^0 = t_{i-1}, t_i^1, ..., t_i^{N-1} = t_i\}$$

be a (possible nonequidistant) partition of every time interval $[t_{i-1}, t_i]$, where $i = 1, ..., r$ such that

$$\max_{0 \leq j \leq N-1} |t_i^{j+1} - t_i^j| \leq \xi_i^N.$$

and $\lim_{N\to\infty} \xi_i^N = 0$ for every $i = 1, ..., r$. Define $\Delta_i t^{j+1} := t_i^{j+1} - t_i^j$, $j = 0, ..., N-1$ and consider the corresponding finite-dimensional optimization problem

$$\begin{aligned} &\text{minimize } J^N(q^N(\cdot), u^N(\cdot)) \text{ and } P^N(q^N(\cdot)), \\ &(q^N(\cdot), u^N(\cdot)) \in (\bigcup_{i=1,...,r} \Gamma_i^N) \times (\mathcal{U}^N \bigcap C^1_{m,N}(0,1)), \end{aligned} \tag{10}$$

where J^N and P^N are discrete variants of the objective functionals J and P from (7). Moreover, Γ_i^N is a correspondingly discrete set Γ_i and $C^1_{m,N}(0,1)$ is set of suitable discrete functions that approximate the trajectories set $C^1_m(0,1)$. Note that the initial continuous optimization problem can also be presented in a similar discrete manner. For example, we can introduce the (Euclidean) spaces of piecewise constant trajectories $q^N(\cdot)$ and piecewise constant control functions $u^N(\cdot)$. As we can see the Banach space $\mathbb{C}^1_n(0,1)$ and the Hilbert space $\mathbb{L}^2_m([0,1])$ will be replaced in that case by some appropriate finite-dimensional spaces.

The discrete optimization problem (10) approximates the infinite-dimensional optimization problem (7). We assume that the set of all weak Pareto optimal solution of the discrete problem (10) is nonempty. Moreover, similarly to the initial optimization problem (7) we also assume that the discrete problem (10) is regular. If $P(\cdot)$ is a convex functional, then the discrete multiobjective optimization problem (10) is also a convex problem. Analogously to the continuous case (7) or (8) we also can write the corresponding KKT optimality conditions for a finite-dimensional optimization problem over the set of variables $(q^N(\cdot), u^N(\cdot))$. The necessary optimality conditions for a discretized problem (10) reduce the finite-dimensional multiobjective optimization problem to a system of nonlinear equations. This problem can be solved by some gradient-based or Newton-type methods (see e.g., [24]).

Finally, note that the proposed numerical approach uses the necessary optimality conditions, namely the KKT conditions, for the discrete variant (10) of the initial optimization problem (7). It is common knowledge that some necessary conditions of optimality for discrete systems, for example the discrete version of the classical Pontryagin Maximum Principle, are non-correct in the absence of some restrictive assumptions. For a constructive numerical treatment of the discrete optimization problem it is necessary to apply some suitable modifications of the conventional optimality conditions. For instance, in the case of discrete optimal control problems one can use so-called Approximate Maximum Principle which is specially designed for discrete approximations of general OCPs [21].

5. Mechanical example

This section is devoted to a short numerical illustration of the proposed hybrid approach to mechanical systems. We deal with a practically motivated model that has the following structure (see Fig. 1).

Let us firstly describe the parameters of the mechanical model under consideration:

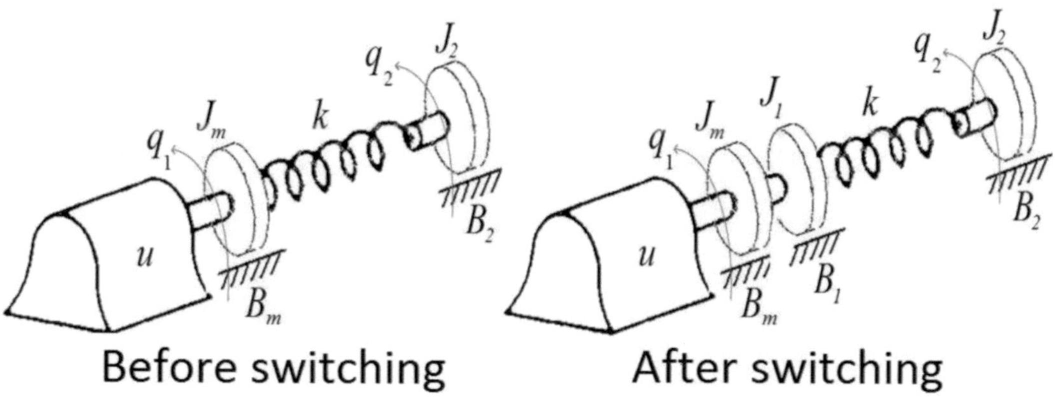

Fig. 1. Mechanical example

- q_1 it corresponds to the position of motor.
- q_2 is the position of inertia J_2.
- J_1, J_2 are the external inertias.
- J_m is an inertia of motor.
- B_m it corresponds to the friction of the motor.
- B_1, B_2 they correspond to the frictions of the inertias J_1, J_2.
- k is a constant called the rate or spring constant.
- u it corresponds to the torque of motor.

The relations for the kinetic potential energies give a rise to the corresponding Lagrange dynamics:

$$K(t) = \frac{1}{2}J_m\dot{q}_1^2 + \frac{1}{2}J_2\dot{q}_2^2$$

$$V(t) = \frac{1}{2}k\left(q_1 - q_2\right)^2$$

Finally, we have

$$L(q(t), \dot{q}(t)) = \frac{1}{2}J_m\dot{q}_1^2 + \frac{1}{2}J_2\dot{q}_2^2 - \frac{1}{2}k\left(q_1 - q_2\right)^2$$

and the Euler-Lagrange equation with respect to the generalized coordinate q_1 has the following form

$$J_m\ddot{q}_1 + B_m\dot{q}_1 - k(q_2(t) - q_1(t)) = u(t) \tag{11}$$

We now considered the Euler-Lagrange equation with respect to the second generalized variable, namely, with respect to q_2

$$\frac{d}{dt}\frac{\partial L(q(t), \dot{q}(t))}{\partial \dot{q}_2} - \frac{\partial L(q(t), \dot{q}(t))}{\partial q_2} = -B_2\dot{q}_2(t)$$

We get the next relation

$$J_2\ddot{q}_2(t) + B_2\dot{q}_2(t) + k(q_2(t) - q_1(t)) = 0$$

The redefinition of the states $x_1 := q_1$, $x_2 := \dot{q}_1$, $x_3 := q_2$, $x_4 := \dot{q}_2$ with $X := (x_1, x_2, x_3, x_4)^T$ implies the compact state-space form of the resulting equation:

$$\dot{X} := \begin{bmatrix} \dot{x}_1 \\ \dot{x}_2 \\ \dot{x}_3 \\ \dot{x}_4 \end{bmatrix} = \begin{bmatrix} 0 & 1 & 0 & 0 \\ \frac{-k}{J_m} & \frac{-B_m}{J_m} & \frac{k}{J_m} & 0 \\ 0 & 0 & 0 & 1 \\ \frac{k}{J_2} & 0 & \frac{-k}{J_2} & \frac{-B_2}{J_2} \end{bmatrix} \begin{bmatrix} x_1 \\ x_2 \\ x_3 \\ x_4 \end{bmatrix} + \begin{bmatrix} 0 \\ \frac{1}{J_m} \\ 0 \\ 0 \end{bmatrix} u, \; X_0 := \begin{bmatrix} x_1^0 \\ x_2^0 \\ x_3^0 \\ x_4^0 \end{bmatrix} \tag{12}$$

The switching structure of the system under consideration is characterized by an additional inertia J_1 and the associated friction B_1. The modified energies are given by the expressions: the kinetic energy:

$$K(t) = \frac{1}{2}J_m\dot{q}_1^2 + \frac{1}{2}J_1\dot{q}_1^2 + \frac{1}{2}J_2\dot{q}_2^2$$

the potential energy:

$$V(t) = \frac{1}{2}k\,(q_1 - q_2)^2$$

The function of Lagrange can be evaluated as follows

$$L(q, \dot{q}) = \frac{1}{2}J_m\dot{q}_1^2 + \frac{1}{2}J_1\dot{q}_1^2 + \frac{1}{2}J_2\dot{q}_2^2 - \frac{1}{2}k\,(q_1 - q_2)^2 \tag{13}$$

The resulting Euler-Lagrange equations (with respect to q_1 and to q_2 can be rewritten as

$$\begin{aligned} (J_m + J_1)\ddot{q}_1(t) + (B_m + B_1)\dot{q}_1(t) - k(q_2(t) - q_1(t)) &= u(t) \\ J_2\ddot{q}_2(t) + B_2\dot{q}_2(t) + k(q_2(t) - q_1(t)) &= 0 \end{aligned} \tag{14}$$

Using the notation introduced above, we obtain the final state-space representation of the hybrid dynamics associated with the given mechanical model:

$$\dot{X} := \begin{bmatrix} \dot{x}_1 \\ \dot{x}_2 \\ \dot{x}_3 \\ \dot{x}_4 \end{bmatrix} = \begin{bmatrix} 0 & 1 & 0 & 0 \\ \frac{-k}{J_m+J_1} & \frac{-(B_m+B_1)}{J_m+J_1} & \frac{k}{J_m+J_1} & 0 \\ 0 & 0 & 0 & 1 \\ \frac{k}{J_2} & 0 & \frac{-k}{J_2} & \frac{-B_2}{J_2} \end{bmatrix} \begin{bmatrix} x_1 \\ x_2 \\ x_3 \\ x_4 \end{bmatrix} + \begin{bmatrix} 0 \\ \frac{1}{J_m+J_1} \\ 0 \\ 0 \end{bmatrix} u \tag{15}$$

The considered mechanical system has a switched nature with a state-dependent switching signal. We put $x_4 = -10$ for the switching-level related to the additional inertia in the system (see above).

Our aim is to find an admissible control law that minimize the value of the quadratic costs functional

$$I(u(\cdot)) = \frac{1}{2}\int_{t_0}^{t_f} \left[X^T(t)QX(t) + u^T(t)Ru(t)\right] dt \longrightarrow \min_{u(\cdot)} \tag{16}$$

The resulting Linear Quadratic Regulator that has the follow form

$$u^{opt}(t) = -R^{-1}(t)B^T(t)P(t)X^{opt}(t) \tag{17}$$

where $P(t)$ is a solution of the Riccati equation (see [7] for details)

$$\dot{P}(t) = -(A^T(t)P(t) + P(t)A(t)) + P(t)B(t)R^{-1}(t)B^T(t)P(t) - Q(t) \tag{18}$$

with the final condition

$$P(t_f) = 0 \tag{19}$$

Let us now present a conceptual algorithm for a concrete computation of the optimal pair $(u^{opt}, X^{opt}(\cdot))$ in this mechanical example. We refer to [7, 8] for the necessary facts and the general mathematical tool related to the hybrid LQ-techniques.

Algorithm 1. *The conceptual algorithm used:*

(0) Select a $t_{swi} \in \left[0, t_f\right]$, *put an index* $j = 0$

(1) Solve the Riccati euqation (18) for (15) on the time intervals $[0, t_{swi}] \cup \left[t_{swi}, t_f\right]$

(2) solve the initial problem (12) for (17)

(3) calculate $x_4(t_{swi}) + 10$, *if* $\mid x_4(t_{swi}) + 10 \mid \cong \epsilon$ *for a prescribed accuracy* $\epsilon > 0$ *then Stop. Else, increase* $j = j + 1$, *inprove* $t_{swi} = t_{swi} + \Delta t$ *and back to (1)*

(4) Finally, solve (15) with the obtained initial conditions(the final conditions for the vector $X(t_{swi})$*) computed from (12)*

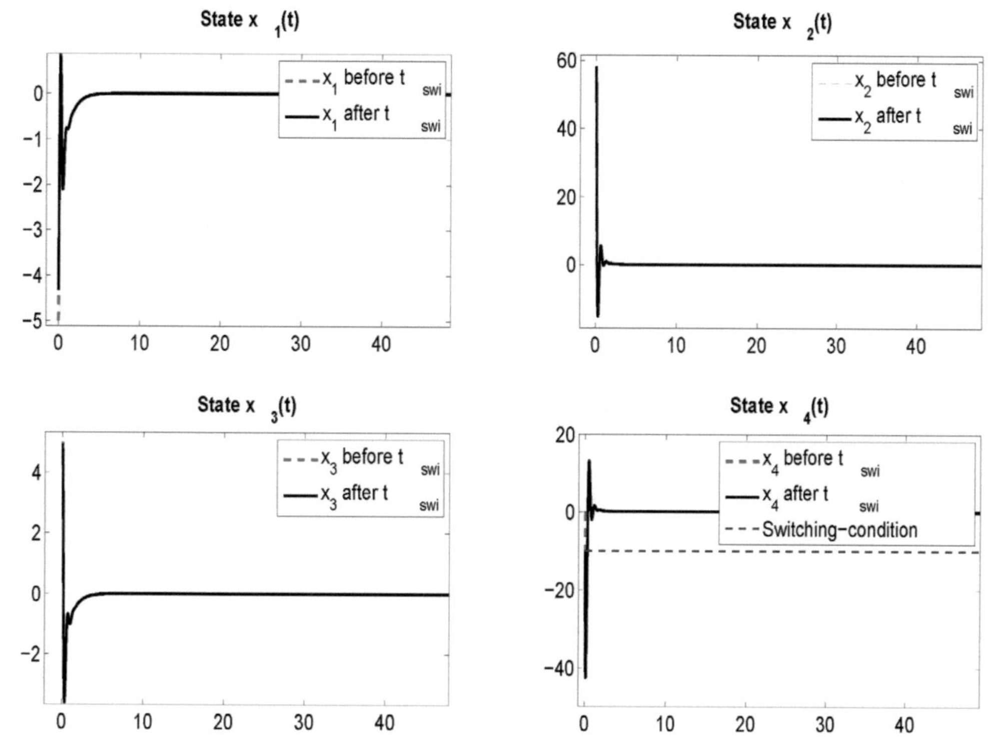

Fig. 2. Components of the optimal trajectories

Finally, let us present the simulation results (figure 2). As we can see, the state x_4 satisfies the switching condition $x_4 + 10 = 0$. The computed switching time is equal to $t_{swi} = 0.0057s$. The dynamic behaviour on the second time-interval $[0, 50]$ is presented on the figure (2). The obtained trajectories of the hybrid states converges to zero. As we can see the dynamic behaviour of the state vector $X^{opt}(t)$ generated by the optimal hybrid control $u^{opt}(\cdot)$ guarantee a minimal value of the quadratic functional $I(\cdot)$. This minimal value characterize the specific control design that guarantee an optimal operation (in the sense of the selected objective) of the hybrid dynamic system under consideration.

6. Concluding remarks

In this paper we propose new theoretical and computational approaches to a specific class of hybrid OCPs motivated by general mechanical systems. Using a variational structure of the nonlinear mechanical systems described by hybrid-type Euler-lagrange or Hamilton equations, one can formulate an auxiliary problem of multiobjective optimization. This problem and the corresponding theoretical and numerical techniques from multiobjective optimization can be effectively applied to numerical solution of the initial hybrid OCP.

The proofs of our results and the consideration of the main numerical concepts are realized under some differentiability conditions and convexity assumptions. These restrictive smoothness assumptions are motivated by the "classical" structure of the mechanical hybrid systems under consideration. On the other hand, the modern variational analysis proceeds without the above restrictive smoothness assumptions. We refer to [20,21] for theoretical details. Evidently, the nonsmooth variational analysis and the corresponding optimization techniques can be considered as a possible mathematical tool for the analysis of discontinuous (for example, variable structure) and impulsive (nonsmooth) hybrid mechanical systems.

Finally, note that the theoretical approach and the conceptual numerical aspects presented in this paper can be extended to some constrained OCPs with additional state and/or mixed constraints. In this case one needs to choose a suitable discretization procedure for the sophisticated initial OCP and to use the corresponding necessary optimality conditions. It seems also be possible to apply our theoretical and computational schemes to some practically motivated nonlinear hybrid and switched OCPs in mechanics, for example, to optimization problems in robots dynamics.

7. References

[1] R. Abraham, *Foundations of Mechanics*, WA Benjamin, New York, 1967.

[2] C.D. Aliprantis and K.C. Border, *Infinite-Dimensional Analysis*, Springer, Berlin, 1999.

[3] V. Azhmyakov and J. Raisch, A gradient-based approach to a class of hybrid optimal control problems, in: *Proceedings of the 2nd IFAC Conference on Analysis and Design of Hybrid Systems*, Alghero, Italy, 2006, pp. 89 – 94.

[4] V. Azhmyakov, S.A. Attia and J. Raisch, On the Maximum Principle for impulsive hybrid systems, *Lecture Notes in Computer Science*, vol. 4981, Springer, Berlin, 2008, pp. 30 – 42.

[5] V. Azhmyakov, An approach to controlled mechanical systems based on the multiobjective technique, *Journal of Industrial and Management Optimization*, vol. 4, 2008, pp. 697 – 712

[6] V. Azhmyakov, V.G. Boltyanski and A. Poznyak, Optimal control of impulsive hybrid systems, *Nonlinear Analysis: Hybrid Systems*, vol. 2, 2008, pp. 1089 – 1097.

[7] V. Azhmyakov, R. Galvan-Guerra and M. Egerstedt, Hybrid LQ-optimization using Dynamic Programming, in: *Proceedings of the 2009 American Control Conference*, St. Louis, USA, 2009, pp. 3617 – 3623.

[8] V. Azhmyakov, R. Galvan-Guerra and M. Egerstedt, Linear-quadratic optimal control of hybrid systems: impulsive and non-impulsive models, *Automatica*, to appear in 2010.

[9] J. Baillieul, The geometry of controlled mechanical systems, in *Mathematical Control Theory* (eds. J. Baillieul and J.C. Willems), Springer, New York, 1999, pp. 322 – 354.

[10] A.M. Bloch and P.E. Crouch, Optimal control, optimization and analytical mechanics, in *Mathematical Control Theory* (eds. J. Baillieul and J.C. Willems), Springer, New York, 1999, pp. 268–321.

[11] M.S. Branicky, S.M. Phillips and W. Zhang, Stability of networked control systems: explicit analysis of delay, in: *Proceedings of the 2000 American Control Conference*, Chicago, USA, 2000, pp. 2352 – 2357.

[12] A. Bressan, Impulsive control systems, in *Nonsmooth Analysis and Geometric Methods in Deterministic Optimal Control*, (eds. B. Mordukhovich and H.J. Sussmann), Springer, New York, 1996, pp. 1 – 22.

[13] C. Cassandras, D.L. Pepyne and Y. Wardi, Optimal control of class of hybrid systems, *IEEE Transactions on Automatic Control*, vol. 46, no. 3, 2001, pp. 398 – 415.

[14] P.D. Christofides and N.H. El-Farra, Control of Nonlinear and Hybrid Processes, *Lecture notes in Control and Information Sciences*, vol. 324, Springer, Berlin, 2005.

[15] F.H. Clarke, *Optimization and Nonsmooth Analysis*, SIAM, Philadelphia, 1990.

[16] G.P. Crespi, I. Ginchev and M. Rocca, Two approaches toward constrained vector optimization and identity of the solutions, *Journal of Industrial and Management Optimization*, vol. 1, 2005, pp. 549 – 563.

[17] M. Egerstedt, Y. Wardi and H. Axelsson, Transition-time optimization for switched-mode dynamical systems, *IEEE Transactions on Automatic Control*, vol. 51, no. 1, 2006, pp. 110 – 115.

[18] F.R. Gantmakher, *Lectures on Analytical Mechanics* (in Russian), Nauka, Moscow, 1966.

[19] D. Liberzon *Switching in Systems and Control*, Birkhäuser, Boston, 2003.

[20] B.S. Mordukhovich, *Variational Analysis and Generalized Differentiation, I: Basic Theory, II Applications*, Springer, New York, 2006.

[21] B.S. Mordukhovich, *Approximation Methods in Problems of Optimization and Optimal Control*, Nauka, Moscow, 1988.

[22] H. Nijmeijer and A.J. Schaft, *Nonlinear Dynamical Control Systems*, Springer, New York, 1990.

[23] B. Piccoli, Hybrid systems and optimal control, in: *Proceedings of the 37th IEEE Conference on Decision and Control*, Tampa, USA, 1998, pp. 13 – 18.

[24] E. Polak, *Optimization*, Springer, New York, 1997.

[25] R. Pytlak, *Numerical Methods for Optimal Control Problems with State Constraints*, Springer, Berlin, 1999.

[26] M.S. Shaikh and P. E. Caines, On the hybrid optimal control problem: theory and algorithms, *IEEE Transactions on Automatic Control*, vol. 52, no. 9, 2007, pp. 1587 – 1603.

[27] H.J. Sussmann, A maximum principle for hybrid optimization, in: *Proceedings of the 38th IEEE Conference on Decision and Control*, Phoenix, USA, 1999, pp. 425 – 430.

[28] Y. Sawaragi, H. Nakayama and T. Tanino, *Theory of Multiobjective Optimization*, Academic Press, Orlando, 1985.
[29] K.L. Teo, C.J. Goh and K.H. Wong, *A Unifed Computational Approach to Optimal Control Problems*, Wiley, New York, 1991.
[30] W. Zhang, M.S. Branicky and S.M. Phillips, Stability of networked control systems, *IEEE Control Systems Magazine*, vol. 21, no. 1, 2004, pp. 84 – 99.

10

Optimized Method for Real Time Nonlinear Control

Younes Rafic[1], Omran Rabih[1] and Rachid Outbib[2]
[1]*Lebanese University, Faculty of Engineering, Beirut*
[2]*LSIS, Aix-Marseilles University, Marseille*
[1]*Lebanon*
[2]*France*

1. Introduction

In this chapter, we discuss the problem of systems control. This problem represents the most important challenge for control engineers. It has attracted the interest of several authors and different approaches have been proposed and tested. These approaches can all be divided into two categories; Linear and Nonlinear approaches. In linear approaches, the analysis and the synthesis are simple however the results are limited to a specified range of operation. In nonlinear approaches, the results are valid in a large domain however the analysis is very complex. We should also note that some works on feedback control are dedicated to the feedback linearization in order to make the models, when it is possible, linear by using a preliminary feedback.

The most important and well-known methodologies about control analysis and feedback control are the following: PID approach, Describing function method, adaptive control, robust control, Lyapunov stability, singular perturbation method, Popov criterion, center manifold theorem and passivity analysis.

The first step in the controller design procedure is the construction of a *truth model* which describes the dynamics of the process to be controlled. The truth model is a *simulation model* that includes the basic characteristics of the process but it is too complicated to be used in the control design. Thus, we need to develop a simplified model to be used instead. Such a model is defined by Friedland (Friedland, 1991) as the *design model*. The design model should capture the essential features of the process.

In order to describe the behavior of the process, a continuous dynamic system constituted by a finite set of ordinary differential equations of the following form is used:

$$\begin{aligned} \dot{x} &= F\left[t, x(t), u(t)\right] \quad x(t_0) = x_0 \\ y(t) &= H\left[t, x(t), u(t)\right] \end{aligned} \tag{1}$$

where the state $x \in \mathbf{R}^n$, the input $u \in \mathbf{R}^m$, the output $y \in \mathbf{R}^p$, and F and H are vector-valued functions with $F : \mathbf{R} \times \mathbf{R}^n \times \mathbf{R}^m \to \mathbf{R}^n$ and $H : \mathbf{R} \times \mathbf{R}^n \times \mathbf{R}^m \to \mathbf{R}^p$.

A second kind of used model is the discrete dynamic system defined by a finite set of difference equations:

$$\begin{aligned} x(k+1) &= F\left[k, x(k), u(k)\right] \quad x(k_0) = x_0 \\ y(k) &= H\left[k, x(k), u(k)\right] \end{aligned} \tag{2}$$

where ***x(k) = x(kh), u(k) = u(kh), h*** is the sampling time interval, and $k \geq 0$ is an integer.

The objective of this chapter is to propose a new strategy for control design using optimization method which is suitable for real time applications. This new methodology is based on neural network which is the classical approach to treat practical results using experimental tests. In order to illustrate this methodology and its applications, we will present an example of the intake air manifold control in a Diesel internal combustion engine.

The chapter is divided as follows: In the second section a short overview of classical control methods is presented. In the third section a new methodology for control is proposed. In the fourth section, we present the application of the new control methodology to the Diesel engine. And finally, we end this chapter with our conclusions and remarks.

2. Overview of classical control methods

A main goal of the feedback control system is to guarantee the stability of the closed-loop behavior. For linear systems, this can be obtained by adapting the control parameters of the transfer function which describes the system in a way so that the real parts of its poles have negative values. Otherwise, Nonlinear control systems use specific theories to ensure the system stability and that is regardless the inner dynamic of the system. The possibility to realize different specifications varies according to the model considered and the control strategy chosen. Hereafter we present a summary of some techniques that can be used:

2.1 Theory of Lyapunov

Lyapunov theory is usually used to determine the stability properties at an equilibrium point without the need to resolve the state equations of the nonlinear system. Let us consider the autonomous non-linear system

$$\dot{x} = F(x) \tag{3}$$

where $x \in R^n$ is the state variable of the system and F is a smooth vector field. Assume that there is a function V defined as follows:

$V : R^n \to R_+$ so that $V(x) = 0 \Leftrightarrow x = 0$ and $\lim\limits_{x \to \infty} V(x) = +\infty$ If the derivative of V along the trajectories of (3) is so that :

$$\dot{V} = \prec \nabla V(x), F(x) \succ \, < 0 \quad \text{for all} \quad x \neq 0 \tag{4}$$

where ∇ designates the gradient and $\prec .,. \succ$ denotes the scalar product, than the system (3) is globally asymptotically stable. This is the Theorem of Lyapunov (Hahn, 1967). This approach has been improved in the principle of Krosoviskii-LaSalle (Hahn, 1967). In fact, it is shown that the condition given by (4) can be relaxed to

$$\dot{V} = \prec \nabla V(x), F(x) \succ \leq 0 \text{ for all } x \tag{5}$$

under the hypothesis that the more invariant set, by (3), included in

$$\Omega = \left\{ x \in R^n \, / \, \dot{V} = 0 \right\} \tag{6}$$

is reduced to the origin.

These two theorems are the base of a large number of results on analysis of stability for nonlinear systems. In fact, the theory of Lyapunov- Krosoviskii-LaSalle is fundamental and is the base of this analysis. In the literature, this theory can have various versions according to the nature of the problem, for instance, for discrete models, stochastic systems or partial differential equations.

In addition to the methodologies developed before, the theory is used to describe the control problems. The use of this theory is illustrated by the following result of feedback stabilization.

Let us consider the following controlled system

$$\dot{x} = F(x) + u \cdot G(x) \tag{7}$$

where $x \in R^n$ is the state, $u \in R$ is the control variable, ***F*** and ***G*** are smooth vector fields. Assume there is V a Lyapunov function so that

$$\dot{V} = \prec \nabla V(x), F(x) \succ \leq 0 \tag{8}$$

Under some hypothesis is proved (Outbib, 1999) that the closed-loop system defined from (7) with

$$u = - \prec \nabla V(x), G(x) \succ \tag{9}$$

is globally asymptotically stable at the origin. A simple example to illustrate this result is the scalar system

$$\dot{x} = u \tag{10}$$

Clearly, the system verifies the hypothesis with $V(x) = 1/2 \cdot x^2$ and the stabilizing control $u = -x$ can be deduced. This approach has been applied to practical process (Outbib, 2000; Dreyfus, 1962)

2.2 Adaptive control

The adaptive control is mainly used in cases where the control law must be continuously adapted due to the varying nature of the system parameters or its initial uncertainties.

Let us consider the following non linear system

$$\dot{x} = F(x, \theta) \tag{11}$$

Where x denotes the state variable of the system and θ designates a parameter. The adaptive control is used in the situation where the parameter θ is not known or can change. For example, let us consider the scalar classical system:

$$\dot{x} = \theta \cdot x^2 + u \tag{12}$$

If θ is known the system (12) can be globally asymptotically stable using a control law of the form $u = -\theta \cdot x^2 - k(x)$, where k is any smooth scalar function defined as follow: $k(x)x > 0$ for $x \neq 0$.

The certainty-equivalent controller is defined by

$$\begin{cases} u = -\hat{\theta} \cdot x^2 - k(x) \\ \dot{\hat{\theta}} = w \end{cases} \tag{13}$$

where w is the update law.
Let V be the Lyapunov function defined by:

$$V(x,\theta) = \frac{1}{2} \cdot x^2 + \frac{\alpha}{2} \cdot \left(\hat{\theta} - \theta\right)^2 \tag{14}$$

with $\alpha > 0$. The derivative of the closed-loop system defined from (12) and (13):

$$\begin{cases} u = -(\theta - \hat{\theta}) \cdot x^2 - k(x) \\ \dot{\hat{\theta}} = w \end{cases} \tag{15}$$

is given by

$$\dot{V}(x,\theta) = -x \cdot k(x) - \left(\hat{\theta} - \theta\right) \cdot x^3 + \alpha \cdot \left(\hat{\theta} - \theta\right) \cdot w \tag{16}$$

Now, if we let $w = (1/\alpha) \cdot x^3$, we get

$$\dot{V}(x,\theta) = -x \cdot k(x) \leq 0 \text{ for all } x \tag{17}$$

This implies that $(x, \hat{\theta})$ is bounded and x converges to zero and ensures that the system (12) can be stabilized at the origin.

2.3 Sliding mode control

The Russian school developed the methodology of sliding mode control in the 1950s. Since this time, the technique has been improved by several authors (Slotine, 1984; Utkin, 1992; Sira-Ramirez, 1987; Bartoloni, 1989; Outbib & Zasadzinski, 2009). This approach is applicable to various domains, including aerospace, robotics, chemical processes, etc.

The sliding mode control is a variable structure control method. Its principle is to force the system to reach and to stay confined over specific surfaces where the stability can be ensured, and that is based on discontinuous control signal.

In order to illustrate the approach based on variable structure control, we now present a simple example. Let us consider the scalar system defined by:

$$\ddot{x} = u \tag{18}$$

Our goal is to propose a control law of the form $u = u(x)$ so that $\lim_{x \to +\infty} x(t) = \lim_{x \to +\infty} \dot{x}(t) = 0$. Clearly, the system (18) can be globally asymptotically stable using a control law of the form $u = f(x, \dot{x})$. In fact, one can use for instance $u = -x - \dot{x}$.

Now a simple analysis can show that there is no linear control law, of the form $u = ax$, which makes the system globally asymptotically stable at the origin. But, if we consider a state feedback that commutes between two linear laws of the form:

$$u = \begin{cases} a_1 \cdot x & if \;\; x \cdot \dot{x} > 0 \\ a_2 \cdot x & if \;\; x \cdot \dot{x} < 0 \end{cases} \tag{19}$$

than the system can be globally asymptotically stable using appropriate values for a_1 and a_2.

2.4 Optimal control

The objective of the optimal control method is to search for the best dynamic course which is capable of transporting the system from an initial state to a final desired state at minimum cost. An example of its various applications can be found in the satellite control. More precisely, the optimal control technique can be defined as follows:

Let us consider the following system:

$$\dot{x} = F(t, x(t), u(t)) \tag{20}$$

where $x \in \mathbf{R}^n$ designates the state variable and $u \in \mathbf{R}^m$ is the control variable. $f : R \times R^n \times R^m \to R^n$ is a smooth vector-valued function .The optimal control is to find a suitable dynamic control *u(t)* which allows the system to follow an optimal trajectory $x(t)$ that minimizes the cost function :

$$J = \int_{t0}^{t1} H(t, x(t), u(t)) \tag{21}$$

Several approaches have been used to resolve this problem. Among these approaches we can cite the variational calculus (Dreyfus, 1962), the maximum principle of Pontryagin (Pontryagin, 1962) or the procedure of dynamic programming method of Bellman (Bellman, 1957).

Let us consider a simple example such as the hanging pendulum. The equation describing the movement of the pendulum angular position under an applied torque α is given by:

$$\begin{cases} \ddot{\theta}(t) + \lambda \cdot \dot{\theta}(t) + \omega^2 \cdot \theta(t) = \alpha(t) \\ \theta(0) = \theta_1 \quad \dot{\theta}(0) = \theta_2 \end{cases} \tag{22}$$

where θ designates the angular position at time t. Let x be the system state variable $x = \left[\theta(t), \dot{\theta}(t)\right]$, we can write :

$$\dot{x}(t) = \begin{pmatrix} x_2 \\ -\lambda \cdot x_2 - \omega^2 \cdot x_1 + \alpha \end{pmatrix} = f(x,\alpha) \tag{23}$$

Therefore the optimal control goal can be to minimize the time interval τ, in order to reach the state values $x(\tau) = 0$.

2.5 Robust control

The objective of robust control is to find a control law in such a way that the response of the system and the error signals are maintained to desired values despite the effects of uncertainties on the system. The uncertainties sources can be any disturbance signals, the measurement noise or the modeling errors due to none considered nonlinearities and time-varying parameters.

The theory of robust control began in the 1970s and 1980s (Doyle, 1979; Zames, 1981) with some aircraft applications. Actually, its applications concern different domains (aerospace, economics, ...).

3. New algorithm for Optimized Nonlinear Control (ONC)

The objective of this methodology is to propose a system optimized dynamic control which can be used in real time control applications. The proposed methodology (Omran, 2008b) can be divided into five steps: 1) Modeling process, 2) Model validation, 3) Dynamic optimization process, 4) Creation of a large database of the optimal control variables using the dynamic optimization process, 5) The neural network controller.

In the next sub-sections, we present the different methodology steps and we explain its application using the example of the Diesel engine system.

3.1 Modeling process

The general equations which describe the functioning of a system can be expressed using the following form (Rivals, 1995):

$$\begin{cases} \dot{X} = F(X,I,u,t) \\ Y = g(X,I,u,t) \end{cases} \tag{24}$$

Where $\boldsymbol{F}$ and g are nonlinear functions, $\boldsymbol{X}$ is the system state variables, $\boldsymbol{I}$ is the inputs variables, $\boldsymbol{u}$ is the control variables to be tuned and Y is the output.

3.2 Experimental validation

In this phase we used specified experimental data to identify the model parameters used in the modeling process (models of representation: transfer function or neural networks, models of knowledge,…), and than we used dynamic experimental data to test the model responses accuracy and its validation. This step is classic in any modeling process.

3.3 Offline dynamic optimization

In this step we present the optimization technique of the control variables over dynamic courses and we define the objective function to be used. The question that we should ask is the following: In response to dynamic inputs I(t) which solicit the system over a certain interval of time [0,T], what is the optimal continuous values of the control parameters u(t) which minimize the cumulative production of the output Y(t). Therefore the objective function to be minimized can be written using the following form:

$$\underset{a_i}{Min}\int_0^T Y(t)\cdot dt = \underset{a_i}{Min}\int_0^T g(X,I,u,t)\cdot dt \tag{25}$$

The optimization problem has the following equalities and inequalities constraints:

Equalities constraints:
$$\frac{dX}{dt} = F(X,I,u,t) \tag{26}$$

Inequalities constraints:
$$\begin{gathered} X_{\min} < X < X_{\max} \\ u_{\min} < u < u_{\max} \end{gathered} \tag{27}$$

Because the problem is nonlinear, there is no analytical solution; therefore we must reformulate it into its discretized form as following:

Objective function:
$$\underset{a_i}{Min}\sum_{i=1}^{N} g_i(X_i,I_i,u_i,t_i) \tag{28}$$

Equality constraints:
$$\frac{dX}{dt} = F(X,I,u,t) \Rightarrow \frac{X_{i+1}-X_i}{\Delta t} = F_i(X_i,I_i,u_i,t_i)$$

$$X_{i+1} - X_i - \Delta t.F(X_i,I_i,u_i,t_i) = 0 \tag{29}$$

Inequality constraints:
$$\begin{gathered} X_{\min} < X_i < X_{\max} \\ u_{\min} < u_i < u_{\max} \end{gathered} \tag{30}$$

The inequality constraints are the domain definition of the system's state and control variables; they are the lower and upper physical, mechanical or tolerance limits which assure a good functioning performance of the system and prevent the system damage. In our case, for example, the engine speed and the intake and exhaust pressure and temperature must vary between a lower and upper limit to prevent engine system damage or dysfunction.

3.4 Creation of the optimal database

The optimization problem explained previously necessitates a long computation time and therefore it cannot be directly resolved in real time applications, in addition the inputs evolution must be known beforehand which is not true in any real time applications. Consequently we propose to resolve the problem off line for different inputs profiles that are very rich in information and variety and that cover a large area of possibility of the system's domain and then to regroup the found solutions (inputs profiles and optimal control variables) in a large database which will be exploited in the following step. Therefore in the created database, we will find for every input vector I(t) an output vector u(t) which is the optimal control variables that can be used to respond to the inputs solicitations. In the next section this database will be used to create a dynamic controller based on neural networks.

3.5 Online neural network control

Since a score of year, the neural networks are considered as a powerful mathematical tool to perform nonlinear regression. Many engineers used them as a black box model to estimate the system responses and they also used them in various fields of applications including pattern recognition, forms recognition, objects classification, filters and control systems (Rivals, 1995). We distinguish two main types of neural networks: feed-forward or multi-layers networks used for steady state processes and feedback or recurrent networks used for dynamic processes. We recognize to these networks the following fundamental characteristics (Ouladssine, 2004): They are black box models with great capacity for universal, flexible and parsimonious functions approximation.

We are interested in establishing a control technique by training a recurrent neural network using the database created in the forth step of this methodology. The main advantage of this approach is essentially the capacity of developing a nonlinear controller with a small computation time which can be executed in real time applications.

Between the various neural networks architectures found in the literature, the multi-layer perceptrons are the most popular; they are particularly exploited in system modeling, identification and control processes (Zweiri 2006). Many works show that the three layers perceptrons with one hidden layer are universal function approximation (Li, 2005); they are capable to approximate any nonlinear continuous function, defined from a finite multi-dimensions space into another, with an arbitrary fixed precision, while they require the identification of a limited number of parameters comparing to the development of series of fixed functions. In this way, they are parsimonious.

4. Application: Optimal air control in diesel engine

Many vehicles developers are especially interested in Diesel internal combustion engines because of their high efficiencies reflecting low fuel consumption. Therefore, electronics and common rail injection systems are largely developed and used in diesel engines along with variable geometry turbocharger and exhaust gas recirculation in order to reduce the pollution and protect the environment and the human health and to optimize the engine performance and fuel consumption. The future engines must respect the more restricted pollution legislations fixed in the European emissions standards (table 1). The particulate matter that are mostly emitted under transient conditions due to air insufficiency are expected to be reduced of a ratio 1:10 at 2010 (Euro 6) and the nitrogen oxides which are caused by a smaller rate of the exhaust gas recirculation due to the insufficiency of fresh air supplied to the engine by the compressor at low engine speed and fuel consumption reduction and engine performance at high speed are also supposed to be reduced to half.

Heavy duty vehicle	Euro 1 1993	Euro 2 1996	Euro 3 2000	Euro 4 2005	Euro 5 2008	Euro 6 2010
Oxides nitrogen	9	7	5	3,5	2	1
Carbon monoxide	4,5	4	2,1	1,5	1,5	1,5
Hydro-carbons	1,23	1,1	0,66	0,46	0,46	0,46
Particulate Matter	0,4	0,15	0,1	0,02	0,02	0,002

Table 1. European standard of heavy duty vehicles in g/KW.h

Actually, modern diesel engines are controlled by look up tables which are the results of a steady state optimization using experiments done on a test bench. Figure 1 shows a static chart of the fresh air flow rate that is used to control the air management system. This chart, as well as the entire look up tables used in the engine control, depends over two entries: the crankshaft angular speed and the fuel mass flow rate (Arnold, 2007). The schematic description of an open and closed loop control are shown in fig. 2 and 3. The inputs are the pedal's position X_p and the engine angular speed w. The outputs are the actuators of the turbine variable geometry *GV* and the opening position of the exhaust gas recirculation EGR. The indication *ref* designates a reference value and the indication *corr* is its corrected value. *P* and m'_C are respectively the predicted or measured intake pressure and the air mass flow rate entering the intake manifold.

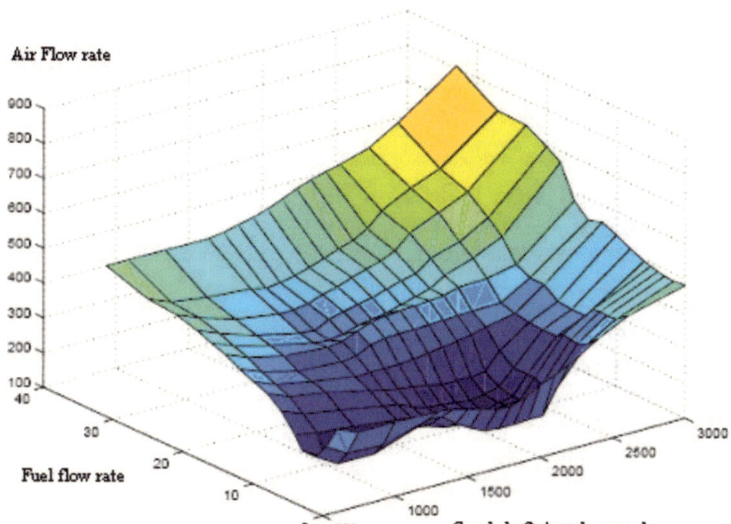

Fig. 1. Static chart of the fresh air flow rate used in the engine control schemes.

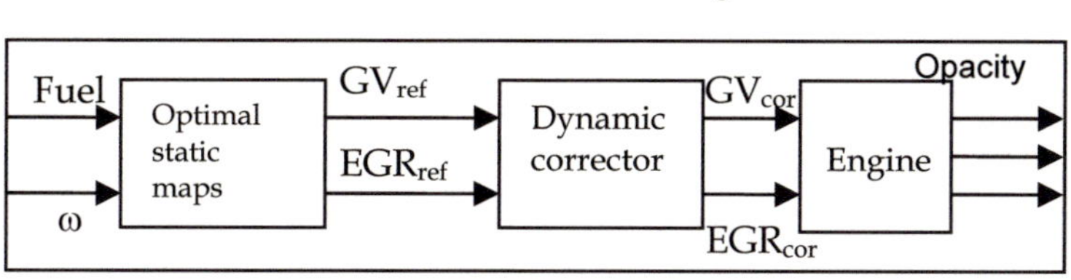

Fig. 2. Open loop control

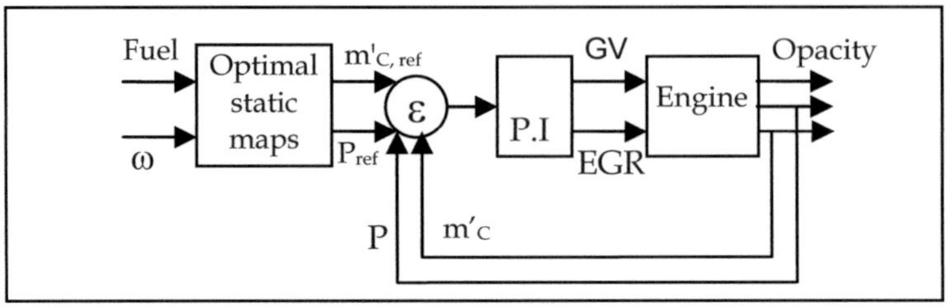

Fig. 3. Closed loop control

In the open loop control, the classic control of a diesel engine (Hafner, 2000) is done according to the diagram in fig. 2, the optimal values of the actuators are updated by memorized static maps. Then a predictive corrector (Hafner, 2001) is generally used in order to compensate the engine dynamic effects.

In the closed loop control (fig. 3), the engine is controlled by error signals which are the difference between the predicted or measured air mass flow rate and the intake pressure, and their reference values. The controller uses memorized maps as reference, based on engine steady state optimization (Hafner, 2001; Bai, 2002). The influence of the dynamic behavior is integrated by several types of controller (PI, robust control with variable parameters, …) (Jung, 2003).

Our work proposes practical solutions to overcome and outperform the control insufficiency using static maps. The advantage of this approach is to be able to propose dynamic maps capable of predicting, "on line", the in-air cylinders filling. Therefore the optimal static maps in fig. 1 and 2 can be replaced by optimal dynamic ones.

We suggest a mathematical optimization process based on the mean value engine model to minimize the total pollutants production and emissions over dynamic courses without deteriorating the engine performance. We used the opacity as a pollution criterion, this choice was strictly limited due to the available data, but the process is universal and it can be applied individually to each pollutant which has physical model or to the all assembled together.

This optimization's procedure is difficult to be applied directly in "on line" engines' applications, due to the computation difficulties which are time consuming. Consequently, it will be used to build up a large database in order to train a neural model which will be used instead. Neural networks are very efficient in learning the nonlinear relations in complex multi-variables systems; they are accurate, easy to train and suitable for real time applications.

All the simulations results and figures presented in this section were computed using Matlab development environment and toolboxes. The following section is divided to four subsections as follows: I Engine dynamic modeling, II Simulation and validation of the engine's model, III Optimization over dynamic trajectories, IV Creation of Neural network for "on line" controller.

4.1 Engine dynamic modeling

Diesel engines can be modeled in two different ways: The models of knowledge quasi-static (Winterbonne, 1984), draining-replenishment (Kao, 1995), semi mixed (Ouenou-Gamo, 2001; Younes, 1993), bond graph (Hassenfolder, 1993), and the models of representation by transfer functions (Younes, 1993), neural networks (Ouladssine, 2004).

Seen our optimization objective, the model of knowledge will be adopted in this work. The semi-mixed model is the simplest analytic approach to be used in an optimization process.

The Diesel engine described here is equipped with a variable geometry turbocharger and water cooled heat exchanger to cool the hot air exiting the compressor, but it doesn't have an exhaust gas recirculation system that is mainly used to reduce the NOx emissions.

Consequently the engine is divided to three main blocks: A. the intake air manifold, B. the engine block, C. the opacity (Omran, 2008a).

4.1.1 Intake air manifold

Considering air as an ideal gas, the state equation and the mass conservation principle gives [4]:

$$V_a \frac{dP_a}{dt} = r.T_a.(\dot{m}_c - \dot{m}_{a0}) \tag{31}$$

$\dot{m}_c$ is the compressor air mass flow rate, $\dot{m}_{a0}$ is the air mass flow rate entering the engine, P_a, V_a and T_a are respectively the pressure, the volume and the temperature of the air in the intake manifold and r is the mass constant of the air. $\dot{m}_{a0}$ is given by:

$$\dot{m}_{a0} = \eta_V \cdot \dot{m}_{a0,th} \tag{32}$$

$\dot{m}_{a0,th}$ is the theoretical air mass flow rate capable of filling the entire cylinders' volume at the intake conditions of pressure and temperature:

$$\dot{m}_{a0,th} = \frac{Vcyl.\omega.P_a}{4 \cdot \pi \cdot r \cdot T_a} \tag{33}$$

V_{cyl} is the displacement, ω is the crankshaft angular speed, and η_v is the in-air filling efficiency given by:

$$\eta_v = a_0 + a_1\omega + a_2\omega^2 \tag{34}$$

Where a_i are constants identified from experimental data. The intake temperature T_a is expressed by:

$$T_a = (1 - \eta_{ech}) \cdot T_c + \eta_{ech} \cdot T_{water} \tag{35}$$

T_c is the temperature of the air at the compressor's exit. T_{water} is the temperature of the cooling water supposed constant. η_{ech} is the efficiency of the heat exchanger supposed constant. The temperature T_c is expressed by:

$$T_c = T_0 \left(1 + \left(\left(\frac{P_a}{P_0} \right)^{\gamma - 1/\gamma} - 1 \right) \frac{1}{\eta_c} \right) \tag{36}$$

4.1.2 Engine block

The principle of the conservation of energy applied to the crankshaft gives:

$$\frac{d}{dt}\left(\frac{1}{2}J(\theta)\omega^2\right) = \mathrm{P}_e - \mathrm{P}_r \tag{37}$$

$J(\theta)$ is the moment of inertia of the engine, it is a periodic function of the crankshaft angle due to the repeated motion of its pistons and connecting rods, but for simplicity, in this paper, the inertia is considered constant. $\mathrm{P_e}$ is the effective power produced by the combustion process:

$$\mathrm{P}_e = \eta_e.\dot{m}_f.P_{ci} \tag{38}$$

$\dot{m}_f$ is the fuel flow rate, P_{ci} is the lower calorific power of fuel and η_e is the effective efficiency of the engine modeled by [5]:

$$\eta_e = \lambda \cdot \begin{pmatrix} c_1 + c_2 \cdot \lambda + c_3 \cdot \lambda^2 + c_4 \cdot \lambda \cdot w + \\ c_5 \cdot \lambda^2 \cdot w + c_6 \cdot \lambda \cdot w^2 + c_7 \cdot \lambda^2 \cdot w^2 \end{pmatrix} \tag{39}$$

c_i are constants, and λ is the coefficient of air excess:

$$\lambda = \frac{\dot{m}_{a0}}{\dot{m}_f} \tag{40}$$

P_r is the resistant power:

$$\mathrm{P}_r = C_r\omega \tag{41}$$

Cr is the resistant torque. Fig. 4 represents a comparison between the effective efficiency model and the experimental data measured on a test bench. The model results are in good agreement with experimental data.

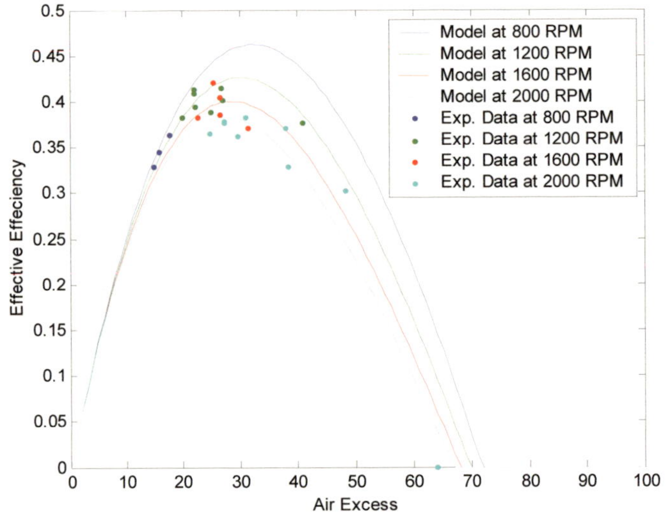

Fig. 4. Comparison between the effective efficiency model results and the experimental data at different crankshaft angular speed.

4.1.3 Diesel emissions model

The pollutants that characterize the Diesel engines are mainly the oxides of nitrogen and the particulate matters. In our work, we are especially interested in the emitted quality of smokes which is expressed by the measure of opacity (Fig. 5) (Ouenou-Gamo, 2001):

$$Opacity = m_1 \cdot w^{m_2} \cdot \dot{m}_a^{m_3 \cdot w + m_4} \cdot \dot{m}_f^{m_5 \cdot w + m_6} \tag{42}$$

m_i are constants identified from the experimental data measured over a test bench.

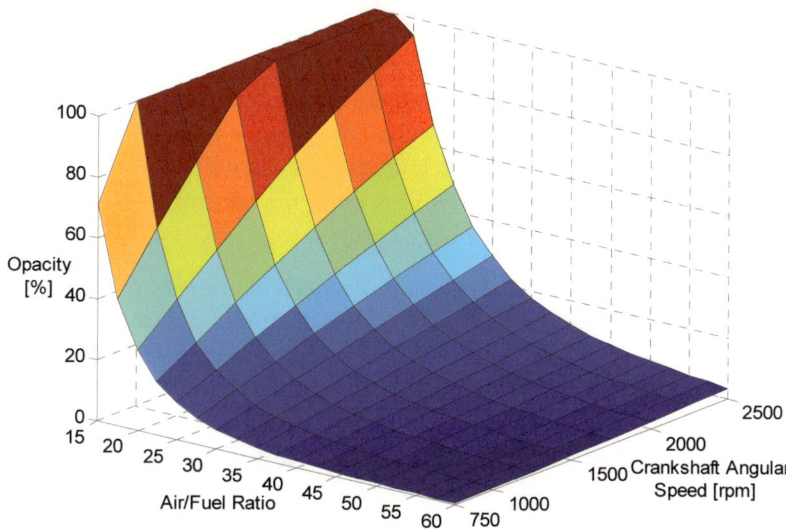

Fig. 5. Graphical representation of the opacity computed using (32) and a constant fuel flow rate equal to 6 g/s.

4.1.4 System complete model

Reassembling the different blocks' equations leads to a complete model describing the functioning and performance of a variable geometry turbocharged Diesel engine. The model is characterized by two state's variables (P_a, w), two inputs ($\dot{m}_f$, C_r) and the following two differential equations representing the dynamic processes:

$$\begin{cases} V_a \cdot \dfrac{dP_a}{dt} = r \cdot T_a \cdot (\dot{m}_c - \dot{m}_{ao}) \\ \dfrac{d}{dt}\left(\dfrac{1}{2} J\omega^2\right) = \eta_e \cdot \dot{m}_f \cdot P_{ci} - C_r \cdot w \end{cases} \tag{43}$$

4.2 Model validation

The test bench, conceived and used for the experimental study, involves: a 6 cylinders turbocharged Diesel engine and a brake controlled by the current of Foucault. Engine's characteristics are reported in table 2.

Stroke [mm]	145
Displacement [cm3]	9839.5
Volumetric ratio	17/1
Bore [mm]	120
Maximum Power [KW] at crankshaft angular speed [rpm]	260 2400
Maximum torque [daN.m] at crankshaft angular speed [rpm]	158 1200
Relative pressure of overfeeding [bar]	2

Table 2. Engine Characteristics

Different systems are used to collect and analyze the experimental data in transient phase and in real time functioning: - Devices for calculating means and instantaneous measures, - a HC analyzer by flame ionization, - a Bosch smoke detector and - an acquisition device for signal sampling. The use of these devices improves significantly the quality of the static measures by integration over a high number of points.

Fig. 6 and 7 show a comparison between two simulations results of the engine complete model and the experimental data. The inputs of the model are the fuel mass flow rate and the resistant torque profiles. The output variables are: the pressure of the intake manifold P_a the crankshaft angular speed ω and the opacity characterizing the engine pollution. The differential equations described in section 4.1.4 are computed simultaneously using the Runge-Kutta method. The simulations are in good agreement with the experimental data.

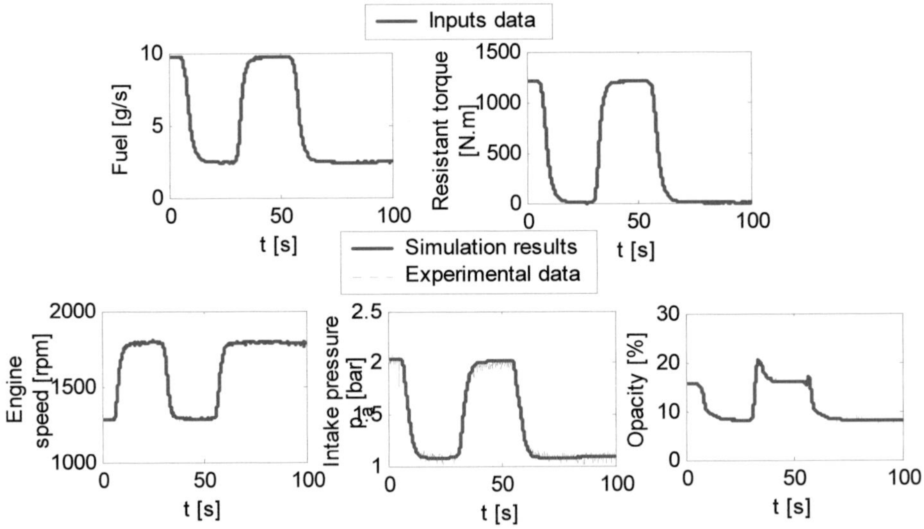

Fig. 6. Simulation 1: Comparison between the complete engine model and the experimental data measured on the test bench.

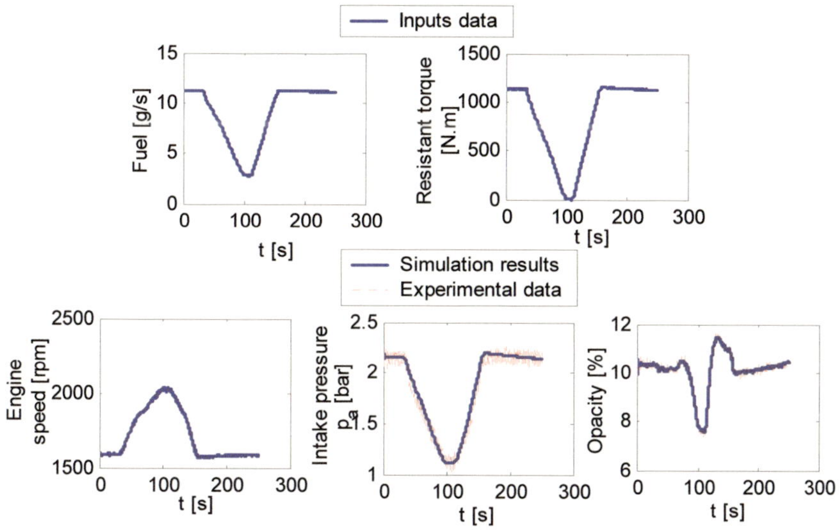

Fig. 7. Simulation 2: Comparison between the complete engine model and the experimental data measured on the test bench.

4.3 Optimization process

4.3.1 Problem description

When conceiving an engine, engines developers have always to confront and solve the contradictory tasks of producing maximum power (or minimum fuel consumption) while respecting several pollution's constraints (European emissions standard). We are only interesting in reducing the pollutants emissions at the engine level, by applying the optimal "in-air cylinders filling". Consequently, the problem can now be defined; it consists in the following objective multi-criteria function:

$$\begin{cases} \text{Maximize "Power"} \\ \text{Minimize "Pollutants"} \end{cases} \tag{44}$$

This multi-objective optimization problem can be replaced by a single, non dimensional, mathematical function regrouping the two previous criteria:

$$f = -\int \frac{\mathrm{P}}{\mathrm{P}_{\max}} \cdot dt + \sum_i \left\{ \int \frac{Poll_i}{Poll_{i,\max}} \cdot dt \right\} \tag{45}$$

P is the engine effective power, $Poll_i$ is a type of pollutant, and the indication *max* characterizes the maximum value that a variable can reach. The integral represents the heap of the pollutants and power over a given dynamic trajectory. This trajectory can be, as an example, a part of the New European Driving Cycle (NEDC).

In this chapter we will only use the opacity as an indication of pollution seen the simplicity of the model and the priority given to the presentation of the method, but we should note that the optimization process is universal and it can involve as many pollution's criteria as we want. The function "objective" becomes:

$$f = -\int \frac{P}{P_{max}} \cdot dt + \int \frac{Op}{Op_{\max}} \cdot dt \tag{46}$$

4.3.2 Formulation of the problem

The problem consists therefore in minimizing the following function "objective" over a definite working interval [0, t]:

$$f = \begin{Bmatrix} -\frac{P_{ci}}{P_{\max}} \cdot \int \eta_e \cdot \dot{m}_f \cdot dt \\ +\frac{m_1}{Op_{\max}} \cdot \int w^{m_2} \cdot \dot{m}_a^{m_3 \cdot w + m_4} \cdot \dot{m}_f^{m_5 \cdot w + m_6} \cdot dt \end{Bmatrix} \tag{47}$$

Under the equalities constraints representing the differential equations of the engine block and the intake manifold:

$$\begin{cases} V_a \cdot \frac{dP_a}{dt} = r \cdot T_a \cdot (\dot{m}_c - \dot{m}_{ao}) \\ \frac{d}{dt}\left(\frac{1}{2} J\omega^2\right) = \eta_e \cdot \dot{m}_f \cdot P_{ci} - C_r \cdot w \end{cases} \tag{48}$$

And the inequalities constraints derived from the physical and mechanical limits of the air excess ratio, the intake pressure and the crankshaft angular speed:

$$\begin{cases} 15 \le \lambda \le 80 \\ 9.5.10^4 \le P_a \le 30.10^4 \quad [Pa] \\ 83 \le \omega \le 230 \quad [rd/s] \end{cases} \tag{49}$$

λ is given by:

$$\lambda = \frac{\left(a_0 + a_1\omega + a_2\omega^2\right) \cdot Ncyl.Vcyl.\omega.P_a}{4 \cdot \pi \cdot \dot{m}_f} \tag{50}$$

The variables of the optimization's problem are w, P_a and m'_c, and the inputs are C_r and m'_f.

We should note that we intentionally eliminate the exhaust and turbo-compressor blocks from the equalities constraints because we are interesting in obtaining the optimal "in-air cylinders filling" m'_c without being limited to any equipments such as the variable geometry turbo-compressor early described. In other words, we can consider that we have replaced the turbo-compressor by a special instrument that can deliver to the intake manifold any value of the air mass flow rate that we choose and at any time. Later, in the conclusion, the devices that can provide these optimal values are briefly discussed. Also we should note that we will use the complete engine model of the existing turbocharged diesel engine as a comparison tool, to prove the validity of our proposed optimal control and the gain in the opacity reduction.

4.3.3 Problem discretization

There is no analytic solution to the problem previously formulated; therefore there is a necessity to reformulate it in its discretized form. The integrals in the function "objective" become a simple sum of the functions computed at different instant t_i:

$$f = \sum_{i=1}^{N} f_i = f_1 + f_2 + \cdots f_N \tag{51}$$

N is the number of the discretized points, it is the size of the unknown vectors $\vec{\omega}$, $\vec{P}_a$ and $\vec{\dot{m}}_c$. h is the step of discretization. Using the Taylor development truncated at the first differential order, the equalities constraints become:

$$\begin{cases} P_{a(i+1)} - P_{a(i)} - \dfrac{h}{V_a}\left(r.T_a.\left(\dot{m}_{c(i)} - \dot{m}_{ao(i)}\right)\right) = 0 \\ \omega^2_{(i+1)} - \omega^2_{(i)} - \dfrac{2 \cdot h}{J}\left(\mathrm{P}_{e(i)} - \mathrm{P}_{r(i)}\right) = 0 \end{cases} \tag{52}$$

And the inequalities constraints:

$$\begin{cases} 15 \le \lambda_{(i)} \le 80 \\ 9.5.10^4 \le P_{a(i)} \le 30.10^4 \quad [Pa] \\ 83 \le \omega_{(i)} \le 230 \quad [rd/s] \end{cases} \tag{53}$$

4.3.4 Solution of the optimization problem

The optimization problem under equality and inequality constraints can be described using the following mathematical form:

$$\begin{cases} Min\{f(X)\} \\ X = (x_1, x_2, \ldots x_n) \\ Under\ Constraints \\ h_i(X) = 0 \quad \mathrm{i} = 1,\ldots,\mathrm{m} \\ g_i(X) \le 0 \quad \mathrm{i} = 1,\ldots,\mathrm{p} \end{cases} \tag{54}$$

Where *f(X)* is the objective function, *h(X)* the equality constraints and *g(X)* the inequality constraints. The easiest way to resolve this problem is to reduce it to a problem without constraints by creating a global objective function *Φ(X, h(X), g(X))* which regroups the original objective function and the equality and inequality constraints (Minoux, 1983).

Therefore we will use a global objective function that regroups: The function "objective", the equalities constraints with Lagrange multipliers, and the inequalities constraints with a penalty function. The final objective function becomes (Minoux, 1983):

$$L(X, \lambda) = f(X) + \sum_{i=1}^{m} \lambda_i * h_i(X) + r.\sum_{i=1}^{p} [g_i(X)]^2 \tag{55}$$

$r = r_0^k$, k is the number of iteration that must tend toward the infinity, and r_0 = 3. The problem will have m additional unknown variables (Lagrange's multipliers λ_i) to be determined along with the engine's variables. The algorithm of the minimization process adopted here is the Broyden-Fletcher-Goldfarb-Shanno B.F.G.S. that sums up as follows:

1. To start by an initial solution $X^{(0)}$.
2. To estimate the solution at the k iteration by: $X_{k+1} = X_k - \alpha_k \cdot D_k \cdot \nabla f(X_k)$, with X is a vector regrouping the optimization variables, a_k is a relaxation factor, $\{D_k \cdot \nabla f(X_k)\}$ represents the decreasing direction of the function, D_k^{-1} is an approximation of the Hessian matrix.
3. To verify if the gradient's module of the objective function at the new vector X is under a certain desired value ($\approx 10^{-2}$). If it is true then this solution is the optimal solution, end of search. Otherwise increment k and return to the stage 2.

4.3.5 Results and discussion

We applied the optimization process explained in the previous section to two different profiles of inputs variables (fuel mass flow rate and resistant torque). The time step of discretization h is equal to 0.01s and the time interval is equal to 3 sec, each problem has 900 unknown variables $\{w,\ P_a \text{ and } \dot{m}_c\}$ with 598 equalities constraints and 1800 inequalities constraints. Fig. 8 and 9 show a comparison between the results of the optimization process and the simulation's results of the engine's complete system model for the same input values and at fixed position of the turbine variable geometry (completely open, GV = 0). The optimization's results show that we need significantly more air mass flow rate entering the intake manifold and higher intake pressure to reduce the opacity, while the real turbocharged diesel engine is not capable of fulfilling these tasks.

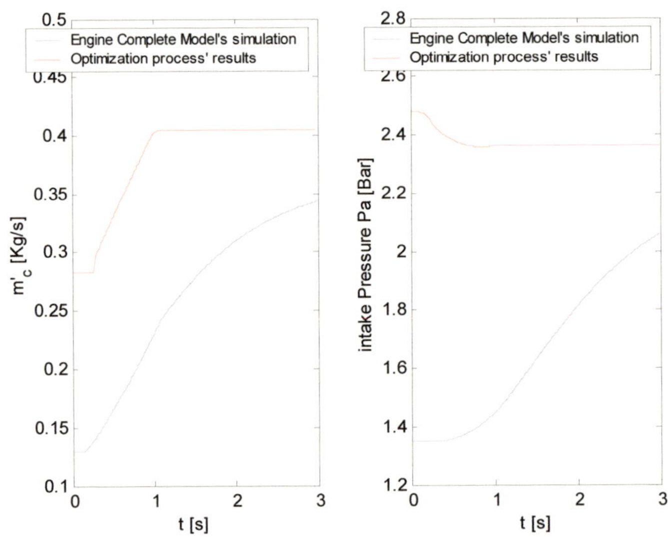

Fig. 8. Comparison between the air mass flow rate and the intake pressure calculated using the optimization procedure and the ones simulated using the engine complete model for a variable fuel flow rate and a constant resistant torque equal to 1000 N.m.

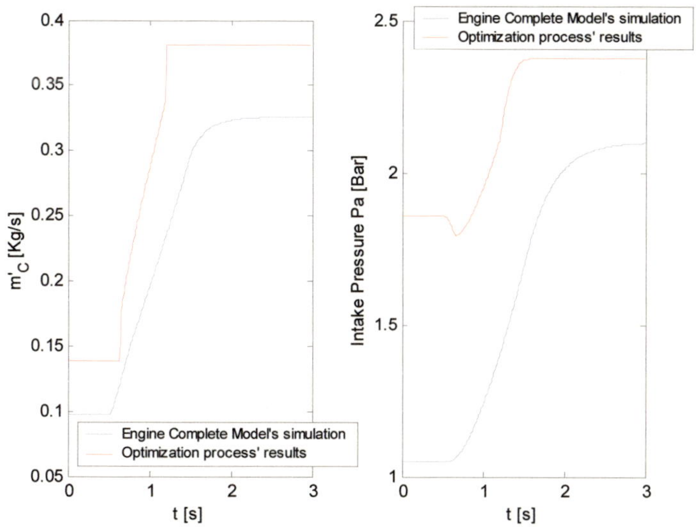

Fig. 9. Comparison between the air mass flow rate and the intake pressure calculated using the optimization procedure and the ones simulated using the engine's complete model for a variable fuel flow rate and a variable resistant torque.

Fig. 10 and 11 show a comparison between the simulated opacity derived from the optimization process and the one derived from the engine's complete system model for the same inputs used in fig. 8 and 9. The enormous gain in the opacity reduction proves the validity of the suggested optimization procedure.

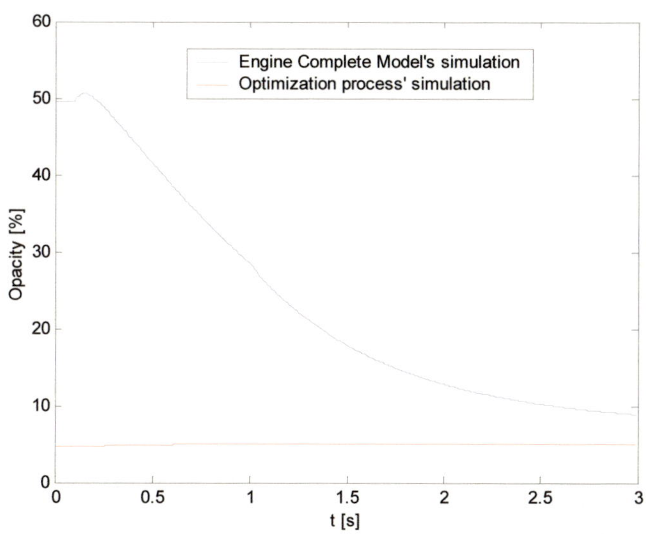

Fig. 10. Opacity reduction using the optimal values of the air mass flow rate and the intake pressure. Blue: simulation without optimization, red: Simulation with optimization.

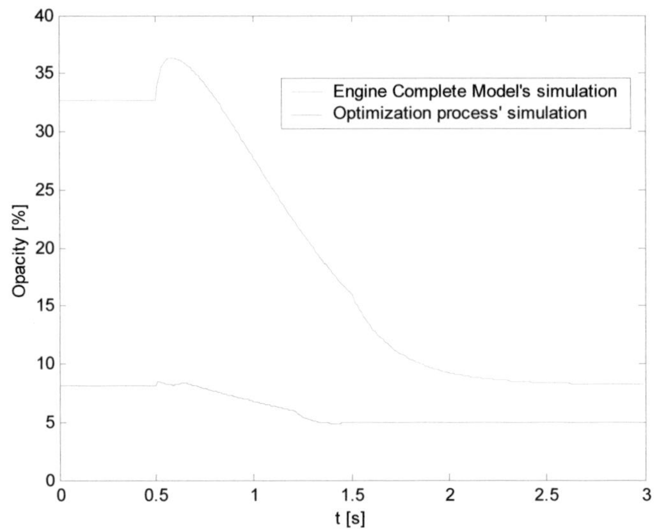

Fig. 11. Opacity reduction using the optimal values of the air mass flow rate and the intake pressure. Blue: simulation without optimization, red: Simulation with optimization.

4.4 Neural network controller

Optimization previously done "off line", would be directly unexploited "on-line" by a controlling processor seen the enormous computation time that is necessary to resolve the optimization problem. In order to integrate the results of this optimization's procedure in a closed loop controller (ref fig. 3), and to be able to use it in real time engine applications, we suggest to use a black box model based on neurons. Neural network is a powerful tool capable of simulating the engine's optimal control variables with good precision and almost instantly.

The neural network inputs are the fuel mass flow rate and the resistant torque, and its output variables are the optimal values of the air mass flow rate and the intake pressure. However in real time engine applications, the injected fuel flow rate is measurable, while the resistant torque is not. Consequently, we suggest substituting this variable by the crankshaft angular speed which can be easily measured and which is widely used in passenger cars controlling systems.

Firstly, we need to create a large database which will be used to train the neural model, and which covers all the functioning area of the engine in order to have a good precision and a highly engine performance. The database is created using the optimization process as explained in subsection 4.3.

Then we have to judicially choose the number of the inputs time sequence to be used, in order to capture the inputs dynamic effects and accurately predict the output variables. With intensive simulations and by trial and error, we find out that a neural network with inputs the fuel mass flow rate and the crankshaft angular speed at instant *(i)*, *(i-1)* and *(i-2)* is capable of precisely predicting the optimal values of the air mass flow rate and the intake pressure at current instant *(i)*. Fig. 12 describes the neural network. The network is built using one hidden layer and one output layer, the activation functions of the hidden layer are sigmoid; the ones at the output layer are linear.

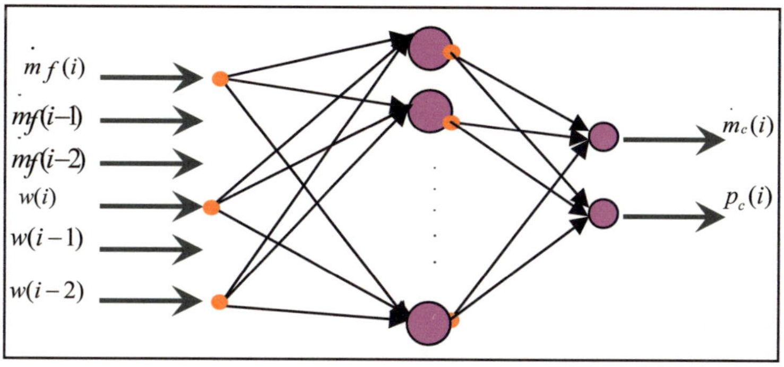

Fig. 12. The structural design of the neural network adopted in this paper for predicting the optimal control of the in-air filling and the intake pressure in real time applications.

The number of neurons in the hidden layer is determined by referring to the errors percentage of the points which are under a certain reference value wisely chosen; the errors percentage (table 3) are the results of the difference between the outputs of the network after the training process is completed, and the desired values used in the training database.

Table 3 shows the results of the neural networks with different number of neurons in their hidden layer, these networks are trained with the same database until a mean relative error equal to 10^{-8} is reached or maximum training time is consumed. The values in the table represent the percentage of the neural network results respecting the specified error percentage computed with respect to the reference values.

Number of neurons of the hidden layer	Error percentage			Relative error
	< 1 %	< 5 %	< 10 %	
110	57.71	88.85	96.71	$3.6\ 10^{-5}$
120	98.428	100	100	10^{-8}
130	98.734	100	100	10^{-8}
140	99	100	100	10^{-8}

Table 3. Results of four neural networks trained using different neurons number in their hidden layer and the same database.

The neural network adopted in this paper includes one hidden layer with 140 neurons and one output layer with 2 neurons. Fig. 13 and 14 show a comparison between the air mass flow rate and the intake pressure calculated using the theoretical optimization procedure, and the ones computed using the neural network. The results are almost identical; the mean relative error is 10^{-6}.

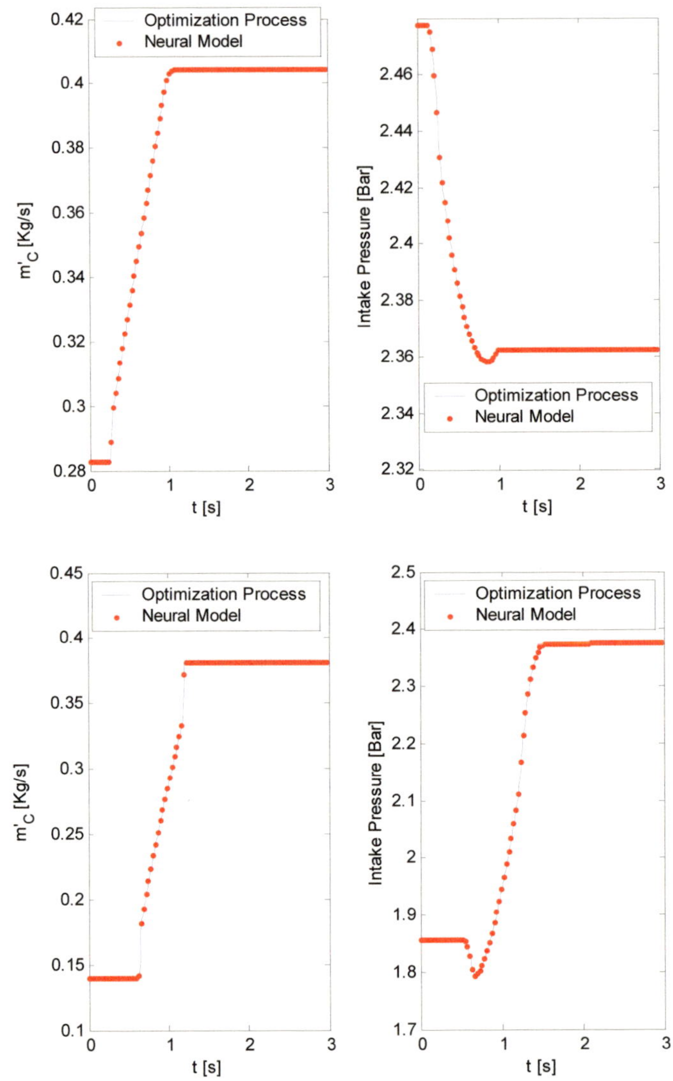

Fig. 13. & 14. Comparison between the neural network outputs and the optimal values of the air mass flow rate and the intake pressure.

5. General conclusions

We successfully developed and validated a mean value physical model that describes the gas states evolution and the opacity of a diesel engine with a variable geometry turbocharger. Then we proposed a dynamic control based on the optimal "in-air cylinders filling" in order to minimize the pollutants emissions while enhancing the engine performance. The optimization process is described in detail and the simulation results (fig. 8-11) prove to be very promising. In addition, the control principle as described here with the opacity criterion can be easily applied to other pollutants which have available physical model. This will be the object of future publications.

Also, in order to overcome on line computation difficulties, a real time dynamic control based on the neural network is suggested; therefore the optimal static maps of the fig. 2 can be successfully replaced by dynamic maps simulated in real time engine functioning (fig. 15).

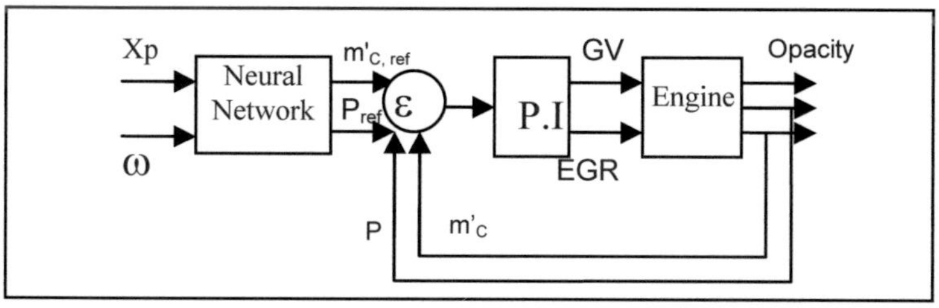

Fig. 15. Proposed control in closed loop

Finally, we should note that, in this chapter, while we did find, in theory, the optimal air mass flow rate and intake pressure necessary to minimize the opacity, but we didn't discuss the mechanical equipments required to provide the optimal intake pressure and intake air flow rate in real time engine applications. The practical implementation of the dynamic control is an important question to be studied thereafter. The use of a turbo-compressor with variable geometry and/or with Waste-Gate, and/or electric compressor is to be considered.

6. References

Arnold, J.F. (2007). *Proposition d'une stratégie de commande à base de logique floue pour la commande du circuit d'air d'un moteur Diesel.* PHD thesis, Rouen University, (March 2007), France.

Bai, L. & Yang, M. (2002). Coordinated control of EGR and VNT in turbocharged Diesel engine based on intake air mass observer, *SAE Technical paper* 2002-01-1292, (March 2002).

Bartoloni, G. (1989). Chattering Phenomena in Discontinuous Control Systems. *International Journal on Systems Sciences*, Vol. 20, Issue 12, (February 1989), ISSN 0020-7721.

Bellman, R. (1975). *Dynamic programming*. Princeton University Press, (1957), Princeton, NJ.

Doyle, J.C. (1979). Robustness of multiloop linear feedback systems" *Proceedings of the 1978 IEEE Conference on Decision and Control*, pages 12–18, Orlando, December 1979.

Dreyfus, S.E. (1962). Variational Problems with Inequality Constraints. *Journal of Mathematical Analysis and Applications*, Vol. 4, Issue 2, (July 1962), ISSN 0022-247X.

Friedland, B. (1996). *Advanced Control System Design*. Prentice-Hall, Englewood Cliffs, ISBN 978-0130140104, NJ, 1996.

Hafner, M. (2000). A Neuro-Fuzzy Based Method for the Design of Combustion Engine Dynamometer Experiments. *SAE Technical Paper* 2000-01-1262, (March 2000).

Hafner, M. (2001). Model based determination of dynamic engine control function parameters. *SAE Technical Paper* 2001-01-1981, (May 2001).

Hahn, W. (1967). *Stability of Motion*, Springer-Verlag, ISBN 978-3540038290, New York, 1967.

Hassenfolder, M. & Gissinger, G.L. (1993). Graphical eider for modelling with bound graphs in processes ». *ICBGM'93*, pp. 188-192, Californie, January 1993.

Jung, M. (2003). *Mean value modelling and robust control of the airpath of a turbocharged diesel engine.* Thesis for doctor of philosophy, University of Cambridge, 2003.

Kao, M. & Moskwa, J.J. (1995). Turbocharger Diesel engine modelling for non linear engine control and state estimation. *Trans ASME, Journal of Dynamic Systems Measurement and Control*, Vol. 117, Issue 1, pp. 20-30, (March 1995), ISSN 0022-0434.

Li, D.; Lu, D.; Kong X. & Wu G. (2005). Implicit curves and surfaces based on BP neural network. *Journal of Information & Computational Science*, Vol. 2, No 2, pp. 259-271, (2005), ISSN 1746-7659.

Minoux, M. (1983). *Programmation Mathématique, Théorie et Algorithmes. tome 1 & 2*, editions dunod, ISBN 978-2743010003, Paris 1983.

Omran, R.; Younes, R. & Champoussin, J.C. (2008a). Optimization of the In-Air Cylinders Filling for Emissions Reduction in Diesel engines, *SAE Technical Paper* 2008-01-1732, (June 2008).

Omran, R. ; Younes, R. & Champoussin, J.C. (2008b). Neural Networks for Real Time non linear Control of a Variable Geometry Turbocharged Diesel Engine, *International Journal of Robust and non Linear Control*, Vol. 18, Issue 12, pp. 1209-1229, (August 2008), ISSN 1099-1239.

Ouenou-Gamo, S. (2001). *Modélisation d'un moteur Diesel suralimenté*, PHD Thesis, Picardie Jules Vernes University, (2001), France.

Ouladssine, M.; Blosh, G. & Dovifazz X. (2004). Neural Modeling and Control of a Diesel Engine with Pollution Constraints. *Journal of Intelligent and Robotic Systems; Theory and Application*, Vol. 41, Issue 2-3, (January 2005), ISSN 0921-0296.

Outbib, R. & Vivalda, J.C. (1999). A note on Feedback stabilization of smooth nonlinear systems. *IEEE Transactions on Automatic Control*. Vol. 44, No 1, pp. 200-203, (August 1999), ISSN 0018-9286.

Outbib, R. & Richard, E. (2000). State Feedback Stabilization of an Electropneumatic System. *ASME Journal of Dynamic Systems Measurement and Control*. Vol. 122, pp 410-415, (September 2000), ISSN 0022-0434.

Outbib, R. & Zasadzinski, M. (2009). Sliding Modes Control, In: *Control Methods for Electrical Machines*, René Husson Editor, Ch. 6, pp. 169-204, 2009. Wiley, ISBN 978-1-84821-093-6, USA.

Pontryagin, L.S.; Boltyanskii, V.; Gamkrelidze, R.V & Mishchenko, E.F. (1962). *The Mathematical Theory of Optimal Processes*. 1962, John Wiley & Sons, USA

Rivals, I. (1995). *Modélisation et commande de processus par réseaux de neurones ; application au pilotage d'un véhicule autonome*, PHD Thesis, Paris 6 University, (1995), France.

Sira-Ramirez, H. (1987). Differential Geometric Methods in Variable-Structure Control. *International Journal of Control*, Vol. 48, Issue 2, (March 1987), pp. 1359-1390, ISSN 0020-7179.

Slotine, J-J.E. (1984). Sliding Controller Design for Non-linear System. *International Journal of Control*, Vol. 40, Issue 2, pp. 421-434, (March 1984), ISSN 0020-7179.

Utkin, V.I. (1992). *Sliding Modes in Control and Optimization*, Springer-Verlag, ISBN 978-0387535166, Berlin 1992.

Winterbonne, D.E. & Horlock, J.H. (1984). *The thermodynamics and gas dynamics internal combustion engines*. Oxford Science Publication, ISBN 978-0198562122, London 1984.

Younes, R. (1993). *Elaboration d'un modèle de connaissance du moteur diesel avec turbocompresseur à géométrie variable en vue de l'optimisation de ses émissions*. PHD Thesis, Ecole Centrale de Lyon, (November 1993), France.

Zames, G. (1981). Feedback and optimal sensitivity: Model reference transformations, multiplicative seminorms, and approximative inverse" *IEEE Transcations on Automatic Control*, Vol. 26, Issue 2, pp. 301–320, (June 1981), ISSN 0018-9286.

Zweiri Y.H. (2006). Diesel engine indicated torque estimation based on artificial neural networks. *International Journal of Intelligent Technology*, Vol. 1, No 1, pp. 233-239, (July 2006), ISSN 1305-6417.

11

Nonlinear Phenomena and Stability Analysis for Discrete Control Systems

Yoshifumi Okuyama
Tottori University, Emeritus
Japan

1. Introduction

Almost all feedback control systems are realized using discretized (discrete-time and discrete-value, i.e., digital) signals. However, the analysis and design method of discretized/quantized (nonlinear) control systems has not been established (Desoer et al., 1975; Elia et al., 2001; Harris et al., 1983; Kalman, 1956; Katz, 1981). This article analyzes the nonlinear phenomena and stability of discretized control systems in a frequency domain[1] (Okuyama, 2006; 2007; 2008). In these studies, it is assumed that the discretization is executed on the input and output sides of a nonlinear element at equal spaces, and the sampling period is chosen of such a size suitable for the discretization in the space. Based on the premise, the discretized (point-to-point) nonlinear characteristic is examined from two viewpoints, i.e., global and local. By partitioning the discretized nonlinear characteristic into two sections and by defining a sectorial area over a specified threshold, the concept of the robust stability condition for nonlinear discrete-time systems is applied to the discretized (hereafter, simply wrriten as discrete) nonlinear control system in question. As a result, the nonlinear phenomena of discrete control systems are clarified, and the stability of discrete nonlinear feedback systems is elucidated.

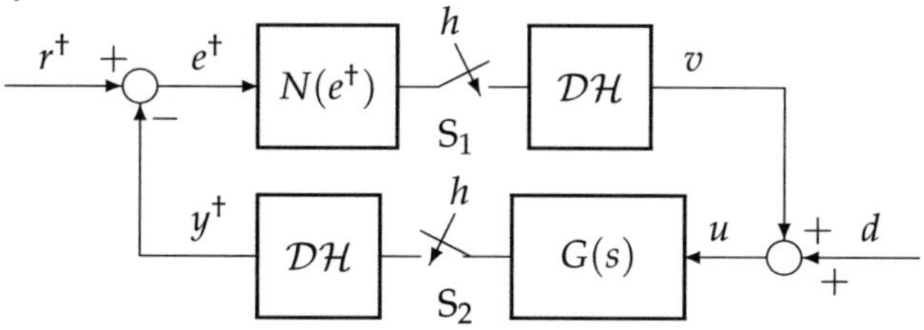

Fig. 1. Nonlinear sampled-data control system.

2. Discrete nonlinear control system

The discrete nonlinear control system to be considered here is represented by a sampled-data control system with two samplers, S_1, S_2 and the continuous nonlinear characteristic $N(\cdot)$ as

[1] In the time domain analysis (e.g., Lyapunov function method), it is difficult to find a Lyapunov function for the discretized (severe nonlinear characteristic) feedback system. The frequency domain analysis will be important in cases where physical systems with uncertainty in the system-order are considered.

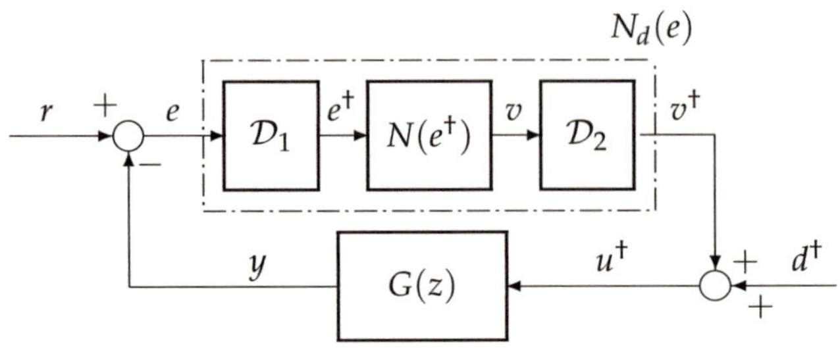

Fig. 2. Discrete nonlinear control system.

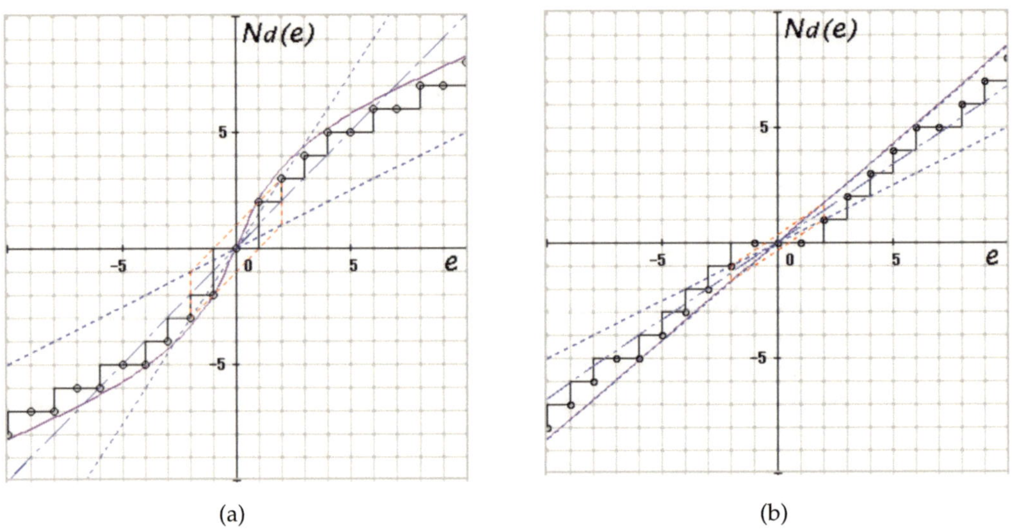

(a)

(b)

Fig. 3. Discretized nonlinear characteristics.

shown in Fig. 1. Here, $\mathcal{DH}$ denotes the discretization and zero-order-hold, which are usually performed in A/D(D/A) conversion, and $G(s)$ is the transfer function of the linear controlled system. It is assumed that the two samplers with a sampling period h operate synchronously.

The sampled-data control system can be equivalently transformed into a discrete control system as shown in Fig. 2. Here, $G(z)$ is the z-transform of $G(s)$ together with zero-order-hold, and $\mathcal{D}_1$ and $\mathcal{D}_2$ are the discretizing units on the input and output sides of the nonlinear element, respectively. The relationship between e and $v^\dagger = N_d(e)$ in the figure becomes a stepwise nonlinear characteristic on integer grid coordinates as shown in Fig. 3 (a). Here, a round-down discretization, which is usually executed on a computer, is applied. Therefore, the relationship between $e^\dagger$ and $u^\dagger$ is indicated by small circles (i.e. a point-to-point transition) on the stepwise nonlinear characteristic. Even if continuous characteristic $N(\cdot)$ is linear, the discretized characteristic $v^\dagger$ becomes nonlinear on integer grid coordinates as shown in Fig. 3 (b) (Okuyama, 2009).

In Fig. 2, each symbol $e, u, y, \cdots$ indicates the sequence $e(k), u(k), y(k), \cdots, (k = 0, 1, 2, \cdots)$ in discrete time, but for continuous value. On the other hand, each symbol $e^\dagger, u^\dagger, \cdots$ indicates a discrete value that can be assigned to an integer number, e.g.,

$$e^{\dagger} \in \{\cdots, -3\gamma, -2\gamma, -\gamma, 0, \gamma, 2\gamma, 3\gamma, \cdots\},$$
$$u^{\dagger} \in \{\cdots, -3\gamma, -2\gamma, -\gamma, 0, \gamma, 2\gamma, 3\gamma, \cdots\},$$

where γ is the resolution of each variable. Here, it is assumed that the input and output signals of the nonlinear characteristic have the same resolution in the discretization. In the figure, $e^{\dagger}$ and $u^{\dagger}$ also represent the sequence $e^{\dagger}(k)$ and $u^{\dagger}(k)$. Without loss of generality, hereafter, $\gamma = 1.0$ is assumed. Thus, the input and output variables of the nonlinear element can be considered in the set of integer numbers, i.e.,

$$e^{\dagger}(k),\ u^{\dagger}(k) \in Z$$
$$Z \stackrel{\text{def}}{=} \{\cdots, -3, -2, -1, 0, 1, 2, 3, \cdots\}.$$

3. Equivalent discrete-time system

In this study, the stepwise and point-to-point nonlinear characteristic is partitioned into the following two sections:

$$N_d(e) = K(e + v(e)),\quad 0 < K < \infty,$$
$$|v(e)| \le \bar{v} < \infty, \tag{1}$$

for $|e| < \varepsilon$, and

$$N_d(e) = K(e + n(e)),\quad 0 < K < \infty,$$
$$|n(e)| \le \alpha|e|,\quad 0 < \alpha \le 1, \tag{2}$$

for $|e| \ge \varepsilon$, where $v(e)$ and $n(e)$ are nonlinear terms relative to nominal linear gain K. Equation (1) represents a bounded nonlinearity which exists in a finite region. On the other hand, (2) represents a sectorial nonlinearity of which the equivalent linear gain exists in a limited range. Therefore, when we consider the robust stability "in a global sense", it is sufficient to consider the nonlinear term $n(e)$. Here, ε is a threshold of the input signal e. As a matter of course, (1) and (2) must be satisfied with respect to the discretized value $e = e^{\dagger}$ because $e^{\dagger} \in e$.

Based on the above consideration, the following new sequences $e_m^{*\dagger}(k)$ and $w_m^{*\dagger}(k)$ are defined:

$$e_m^{*\dagger}(k) = e_m^{\dagger}(k) + q \cdot \frac{\Delta e^{\dagger}(k)}{h}, \tag{3}$$

$$w_m^{*\dagger}(k) = w_m^{\dagger}(k) - \alpha q \cdot \frac{\Delta e^{\dagger}(k)}{h}. \tag{4}$$

where q is a non-negative number, $e_m^{\dagger}(k)$ and $w_m^{\dagger}(k)$ are neutral points of sequences $e^{\dagger}(k)$ and $w^{\dagger}(k)$,

$$e_m^{\dagger}(k) = \frac{e^{\dagger}(k) + e^{\dagger}(k-1)}{2}, \tag{5}$$

$$w_m^{\dagger}(k) = \frac{w^{\dagger}(k) + w^{\dagger}(k-1)}{2}, \tag{6}$$

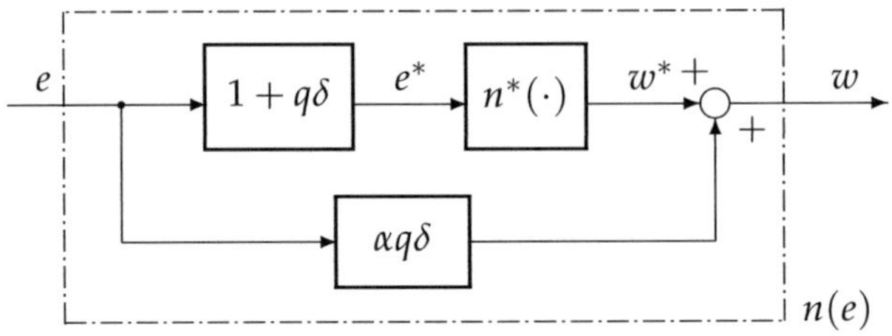

Fig. 4. Nonlinear subsystem.

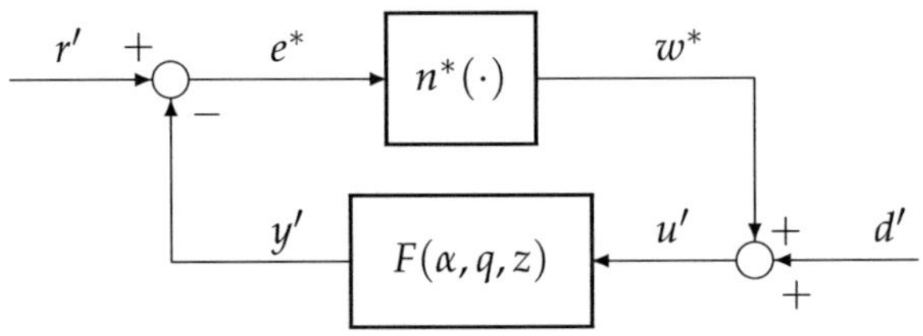

Fig. 5. Equivalent feedback system.

and $\Delta e^{\dagger}(k)$ is the backward difference of sequence $e^{\dagger}(k)$,

$$\Delta e^{\dagger}(k) = e^{\dagger}(k) - e^{\dagger}(k-1). \tag{7}$$

The relationship between equations (3) and (4) in regard to the continuous values is shown by the block diagram in Fig. 4. In this figure, δ is defined as

$$\delta(z) := \frac{2}{h} \cdot \frac{1 - z^{-1}}{1 + z^{-1}}. \tag{8}$$

Equation (8) corresponds to the bilinear transformation between z and δ. Thus, the loop transfer function from w^* to e^* can be given by $F(\alpha, q, z)$, as shown in Fig. 5, where

$$F(\alpha, q, z) = \frac{(1 + q\delta(z))KG(z)}{1 + (1 + \alpha q\delta(z))KG(z)}, \tag{9}$$

and r', d' are transformed exogenous inputs. Here, the variables such as w^*, u' and y' written in Fig. 5 indicate the z-transformed ones.

In this study, the following assumption is provided on the basis of the relatively fast sampling and the slow response of the controlled system.

[Assumption] The absolute value of the backward difference of sequence $e(k)$ is not more than γ, i.e.,

$$|\Delta e(k)| = |e(k) - e(k-1)| \leq \gamma. \tag{10}$$

If the condition (10) is satisfied, $\Delta e^{\dagger}(k)$ defined by (7) is exactly $\pm\gamma$ or 0 because of the discretization. That is, the absolute value of the backward difference can be given as

$$|\Delta e^{\dagger}(k)| = |e^{\dagger}(k) - e^{\dagger}(k-1)| = \gamma \text{ or } 0. \quad \square$$

This assumption will be satisfied in the following examples.

4. Norm conditions

In this section, some lemmas for norm conditions are presented. Here, in regard to (2), the following new nonlinear function is defined.

$$f(e) := n(e) + \alpha e. \tag{11}$$

When considering the discretized output of the nonlinear term, $w^\dagger = n(e^\dagger)$, the following expression can be given:

$$f(e^\dagger(k)) = w^\dagger(k) + \alpha e^\dagger(k). \tag{12}$$

From inequality (2), it can be seen that the function (12) belongs to the first and third quadrants. Considering the equivalent linear characteristic, the following inequality can be defined:

$$0 \leq \beta(k) := \frac{f(e^\dagger(k))}{e^\dagger(k)} \leq 2\alpha. \tag{13}$$

When this type of nonlinearity $\beta(k)$ is used, inequality (2) can be expressed as

$$w^\dagger(k) = n(e^\dagger(k)) = (\beta(k) - \alpha)e^\dagger(k). \tag{14}$$

For the neutral points of $e^\dagger(k)$ and $w^\dagger(k)$, the following expression is given from (12):

$$\frac{1}{2}(f(e^\dagger(k)) + f(e^\dagger(k-1))) = w_m^\dagger(k) + \alpha e_m^\dagger(k). \tag{15}$$

Moreover, equation (14) is rewritten as

$$w_m^\dagger(k) = (\beta(k) - \alpha)e_m^\dagger(k).$$

Since $|e_m^\dagger(k)| \leq |e_m(k)|$, the following inequality is satisfied when a round-down discretization is executed:

$$|w_m^\dagger(k)| \leq \alpha|e_m^\dagger(k)| \leq \alpha|e_m(k)|. \tag{16}$$

Based on the above premise, the following norm inequalities are examined (Okuyama et al., 1999; Okuyama, 2006).

[Lemma-1] The following inequality holds for a positive integer p:

$$\|w_m^\dagger(k)\|_{2,p} \leq \alpha\|e_m^\dagger(k)\|_{2,p} \leq \alpha\|e_m(k)\|_{2,p}. \tag{17}$$

Here, $\|\cdot\|_{2,p}$ denotes the Euclidean norm, which can be defined by

$$\|x(k)\|_{2,p} := \left(\sum_{k=1}^{p} x^2(k)\right)^{1/2}.$$

(Proof) The proof is clear from inequality (16). □

[Lemma-2] If the following inequality is satisfied in regard to the inner product of the neutral points of (12) and the backward difference (7):

$$\langle\, w_m^\dagger(k) + \alpha e_m^\dagger(k), \Delta e^\dagger(k) \,\rangle_p \geq 0, \tag{18}$$

the following inequality can be obtained:

$$\|w_m^{*\dagger}(k)\|_{2,p} \leq \alpha \|e_m^{*\dagger}(k)\|_{2,p} \tag{19}$$

for any $q \geq 0$. Here, $\langle \cdot, \cdot \rangle_p$ denotes the inner product, which can be defined as

$$\langle\, x_1(k), x_2(k)\,\rangle_p = \sum_{k=1}^{p} x_1(k)x_2(k).$$

(Proof) The following equation is obtained from (3) and (4):

$$\alpha^2 \|e_m^{*\dagger}(k)\|_{2,p}^2 - \|w_m^{*\dagger}(k)\|_{2,p}^2 = \alpha^2 \left\| e_m^{\dagger}(k) + q\frac{\Delta e^{\dagger}(k)}{h} \right\|_{2,p}^2 - \left\| w_m^{\dagger}(k) - \alpha q\frac{\Delta e^{\dagger}(k)}{h} \right\|_{2,p}^2$$

$$= \alpha^2 \|e_m^{\dagger}(k)\|_{2,p}^2 - \|w_m^{\dagger}(k)\|_{2,p}^2 + \frac{2\alpha q}{h} \cdot \langle w_m^{\dagger}(k) + \alpha e_m^{\dagger}(k), \Delta e^{\dagger}(k)\rangle_p. \tag{20}$$

Thus, (19) is satisfied by using the left inequality of (17). □

In regard to the input of $n^*(\cdot)$, the following inequality can be obtained from (20) and the second inequality of (17) as follows:

$$\|w_m^{*\dagger}(k)\|_{2,p} \leq \alpha \|e_m^{*}(k)\|_{2,p}, \tag{21}$$

when inequality (18) is satisfied.

5. Sum of trapezoidal areas

The left side of inequality (18) can be expressed as a sum of trapezoidal areas.

[Lemma-3] For any step p, the following equation is satisfied:

$$\sigma(p) := \langle\, w_m^{\dagger}(k) + \alpha e_m^{\dagger}(k), \Delta e^{\dagger}(k)\,\rangle_p = \frac{1}{2}\sum_{k=1}^{p}(f(e^{\dagger}(k)) + f(e^{\dagger}(k-1)))\Delta e^{\dagger}(k). \tag{22}$$

(Proof) The proof is clear from (15). □

In general, the sum of trapezoidal areas holds the following property.
[Lemma-4] If inequality (10) is satisfied in regard to the discretization of the control system, the sum of trapezoidal areas becomes non-negative for any p, that is,

$$\sigma(p) \geq 0. \tag{23}$$

(Proof) Since $f(e^{\dagger}(k))$ belongs to the first and third quadrants, the area of each trapezoid

$$\tau(k) := \frac{1}{2}(f(e^{\dagger}(k)) + f(e^{\dagger}(k-1)))\Delta e^{\dagger}(k) \tag{24}$$

is non-negative when $e(k)$ increases (decreases) in the first (third) quadrant. On the other hand, the trapezoidal area $\tau(k)$ is non-positive when $e(k)$ decreases (increases) in the first (third) quadrant.

Strictly speaking, when ($e(k) \geq 0$ and $\Delta e(k) \geq 0$) or ($e(k) \leq 0$ and $\Delta e(k) \leq 0$), $\tau(k)$ is non-negative for any k. On the other hand, when ($e(k) \geq 0$ and $\Delta e(k) \leq 0$) or ($e(k) \leq$

0 and $\Delta e(k) \geq 0$), $\tau(k)$ is non-positive for any k. Here, $\Delta e(k) \geq 0$ corresponds to $\Delta e^{\dagger}(k) = \gamma$ or 0 (and $\Delta e(k) \leq 0$ corresponds to $\Delta e^{\dagger}(k) = -\gamma$ or 0) for the discretized signal, when inequality (10) is satisfied.

The sum of trapezoidal area is given from (22) as:

$$\sigma(p) = \sum_{k=1}^{p} \tau(k). \tag{25}$$

Therefore, the following result is derived based on the above. The sum of trapezoidal areas becomes non-negative, $\sigma(p) \geq 0$, regardless of whether $e(k)$ (and $e^{\dagger}$(k)) increases or decreases. Since the discretized output traces the same points on the stepwise nonlinear characteristic, the sum of trapezoidal areas is canceled when $e(k)$ (and $e^{\dagger}(k)$ decreases (increases) from a certain point $(e^{\dagger}(k), f(e^{\dagger}(k)))$ in the first (third) quadrant. (Here, without loss of generality, the response of discretized point $(e^{\dagger}(k), f(e^{\dagger}(k)))$ is assumed to commence at the origin.) Thus, the proof is concluded. □

6. Stability in a global sense

By applying a small gain theorem to the loop transfer characteristic (9), the following robust stability condition of the discrete nonlinear control system can be derived.

[Theorem] If there exists a $q \geq 0$ in which the sector parameter α in regard to nonlinear term $n(\cdot)$ satisfies the following inequality, the discrete-time control system with sector nonlinearity (2) is robust stable in an ℓ_2 sense:

$$\alpha < \eta(q,\omega) := \frac{-q\Omega V + \sqrt{q^2\Omega^2V^2 + (U^2+V^2)\{(1+U)^2+V^2\}}}{U^2+V^2}, \tag{26}$$
$$\forall \omega \in [0, \omega_c], \quad \omega_c : \text{ cutoff frequency}$$

when the linearized system with nominal gain K is stable. Here, $\Omega(\omega)$ is the distorted frequency of angular frequency ω and is given by

$$\delta(e^{j\omega h}) = j\Omega(\omega) = j\frac{2}{h}\tan\left(\frac{\omega h}{2}\right), \quad j = \sqrt{-1}. \tag{27}$$

In addition, $U(\omega)$ and $V(\omega)$ are the real and the imaginary parts of $KG(e^{j\omega h})$, respectively.

(Proof) Based on the loop characteristic in Fig. 5, the following inequality can be given in regard to $z = e^{j\omega h}$:

$$\|e_m^*(z)\|_{2,p} \leq c_1\|r_m'(z)\|_{2,p} + c_2\|d_m'(z)\|_{2,p} + \sup_{z=1}|F(\alpha,q,z)| \cdot \|w_m^{*\dagger}(z)\|_{2,p}.$$

Here, $r_m'(z)$ and $d_m'(z)$ denote the z-transformation for the neutral points of sequences $r'(k)$ and $d'(k)$, respectively. Moreover, c_1 and c_2 are positive constants.

By applying inequality (21), the following expression is obtained:

$$\left(1 - \alpha \cdot \sup_{z=1}|F(\alpha,q,z)|\right)\|e_m^*(z)\|_{2,p} \leq c_1\|r_m'(z)\|_{2,p} + c_2\|d_m'(z)\|_{2,p}. \tag{28}$$

Therefore, if the following inequality (i.e., the small gain theorem with respect to ℓ_2 gains) is valid,

$$|F(\alpha, q, e^{j\omega h})| \leq 1/\alpha, \tag{29}$$

the sequences $e_m^*(k)$, $e_m(k)$, $e(k)$ and $y(k)$ in the feedback system are restricted in finite values when exogenous inputs $r(k)$, $d(k)$ are finite and $p \to \infty$.

By substituting (9) into inequality (29), the following is obtained:

$$\left| \frac{(1+jq\Omega(\omega))KG(e^{j\omega h})}{1+(1+j\alpha q\Omega(\omega))KG(e^{j\omega h})} \right| < \frac{1}{\alpha}. \tag{30}$$

From the square of both sides of inequality (30),

$$\alpha^2(1+q^2\Omega^2)(U^2+V^2) < (1+U-\alpha q\Omega V)^2 + (V+\alpha q\Omega U)^2$$

Then,

$$\alpha^2(U^2+V^2) + 2\alpha q\Omega V - \{(1+U^2)+V^2\} < 0. \tag{31}$$

Consequently, as a solution of inequality (31),

$$\alpha < \frac{-q\Omega V + \sqrt{q^2\Omega^2V^2 + (U^2+V^2)\{(1+U)^2+V^2\}}}{U^2+V^2}$$

can be given. □

Since inequality (26) in Theorem-1 is for all ω (and Ω) considered and a certain q, the condition is rewritten as the following max-min problem.

[Corollary] If the following inequality is satisfied, the discrete-time control system with sector nonlinearity (2) is robust stable:

$$\alpha < \eta(q_0, \omega_0) = \max_{q} \min_{\omega} \eta(q, \omega), \tag{32}$$

when the linearized system with nominal gain is stable. □

In this study, a non-conservative sufficient condition for the stability of discrete-time and discrete-value control systems is derived by applying the concept of robust stability in our previous paper(Okuyama et al., 2002a). The stability condition is, however, not satisfied for the entire area of the input of nonlinearity $N(e)$ because of the stepwise and point-to-point characteristic. Even if the response seems to be asymptotic, there may remain a fluctuation (a sustained oscillation in the discrete time) or an offset. Of course, a divergent response that reaches the sustained oscillation may occur. These responses are typical nonlinear phenomena. The theorem (and corollary) derived here should be considered as the robust stability condition in a global sense. In addition, it is valid based on an assumption in the relationship between the sampling period and the system dynamics. However, this result will be useful in designing a discrete (digital, packet transmission) control system in practice.

Naturally, the stability condition becomes that of continuous-time and continuous-value nonlinear control systems, when the sampling period h and the resolution γ approach zero. Inequality (26) in Theorem-1 corresponds to Popov's criterion for discrete-time systems and contains the circle criterion for nonlinear time-varying (discrete-time) systems in a special case. The relationship between them will be described in the next section.

7. Relation to Popov's criterion

Inequality (30) can be rewritten as follows:

$$\left| \frac{\alpha H(\alpha, q, e^{j\omega h})}{1 + \alpha H(\alpha, q, e^{j\omega h})} \right| < 1, \tag{33}$$

where

$$H(\alpha, q, e^{j\omega h}) = \frac{(1 + jq\Omega(\omega))KG(e^{j\omega h})}{1 + (1-\alpha)KG(e^{j\omega h})}.$$

From (33), the following inequality is obtained:

$$2\alpha \cdot \Re\{H(\alpha,, q, e^{j\omega h})\} + 1 > 0. \tag{34}$$

Therefore, the following robust stability condition can be given:

$$\Re\left\{ \frac{1 + (1+\alpha)KG(e^{j\omega h}) + 2j\alpha q\Omega(\omega)KG(e^{j\omega h})}{1 + (1-\alpha)KG(e^{j\omega h})} \right\} > 0, \tag{35}$$

which is equivalent to inequality (26). When $\alpha = 1$ is chosen, (35) can be written as follows:

$$\frac{1}{K_m} + \Re\{(1 + jq\Omega(\omega))G(e^{j\omega h})\}, \tag{36}$$

where $K_m = 2K$. In this case, the allowable sector of nonlinear characteristic $N(\cdot)$ is given as

$$0 \leq N(e)e \leq K_m e^2, \quad e \neq 0. \tag{37}$$

When h approaches zero (or ω is a low frequency), inequalities (36) and (37) are equivalent to an expression of Popov's criterion for continuous-time systems.

In case of $q = 0$, the left side of (26) becomes the inverse the absolute value of complementary sensitivity function $T(j\omega)$.

$$\eta(0, \omega) = \frac{\sqrt{(1 + U^2)) + V^2}}{\sqrt{U^2 + V^2}} = \frac{1}{|T(j\omega)|} > \alpha. \tag{38}$$

On the other hand, from (35)

$$\Re\left\{ \frac{1 + (1+\alpha)KG(e^{j\omega h})}{1 + (1-\alpha)KG((e^{j\omega h})} \right\} > 0 \tag{39}$$

is obtained. Inequalities (38) and (39) correspond to the circle criterion for nonlinear time-varying systems.

8. Validity of Aizerman's conjecture

In the following case, Theorem-1 becomes equal to the robust stability condition of the linear interval gain that corresponds to Aizerman's conjecture which was extended into discrete-time systems (Okuyama et al., 1998).

[Theorem-2] If the right side of (32) is satisfied at the saddle point,

$$\left(\frac{\partial \eta(q, \omega)}{\partial q} \right)_{q=q_0, \omega=\omega_0} = 0, \tag{40}$$

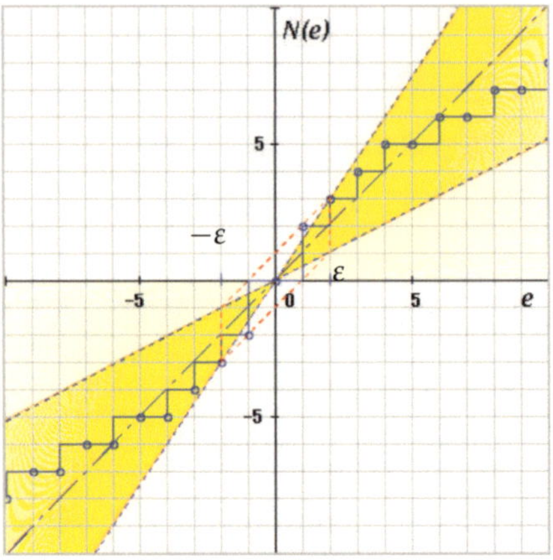

Fig. 6. Discretized nonlinear characteristic and stable sector for Example-1.

inequality (26) of Theorem-1 becomes equal to the robust stability condition provided for a linear time-invariant discrete-time system.
(Proof) This theorem can easily be proven by using the right side of (26). Then,

$$\frac{\partial \eta(q,\omega)}{\partial q} = \frac{-\eta(q,\omega)\Omega(\omega)V(\omega)}{\sqrt{q^2\Omega^2 v^2 + (U^2+V^2)\{(1+U)^2+V^2\}}}. \tag{41}$$

From (40), the following can be obtained:

$$\eta(q_0,\omega_0)\Omega(\omega_0)V(\omega_0) = 0. \tag{42}$$

Obviously, $\eta(q,\omega) > 0$. Moreover, since $0 < \omega_0 < \pi/h, \Omega(\omega_0) > 0$ from (27). Then,

$$V(\omega_0) = 0 \tag{43}$$

is obtained. Thus,

$$\eta(q_0,\omega_0) = \frac{|1+U(\omega_0)|}{|U(\omega_0)|} > \alpha \tag{44}$$

Inequality (44) corresponds to the stability condition which was determined for the time-invariant discrete-time system with a linear gain, i.e., the Nyquist stability condition for a discrete-time system.

Theorem-2 shows that the robust stability condition for a linear time-invariant system (the concept of interval set) can be applied to nonlinear discrete-time control systems, when (40) is satisfied. However, (32) is not always valid at the saddle point given in (40). In the following example, it can be shown that there are counter examples of Aizerman's conjecture extended into the nonlinear discrete-time systems.

9. Numerical examples

In order to verify the theoretical result, two numerical examples for discrete control systems with saturation type nonlinearity are presented.

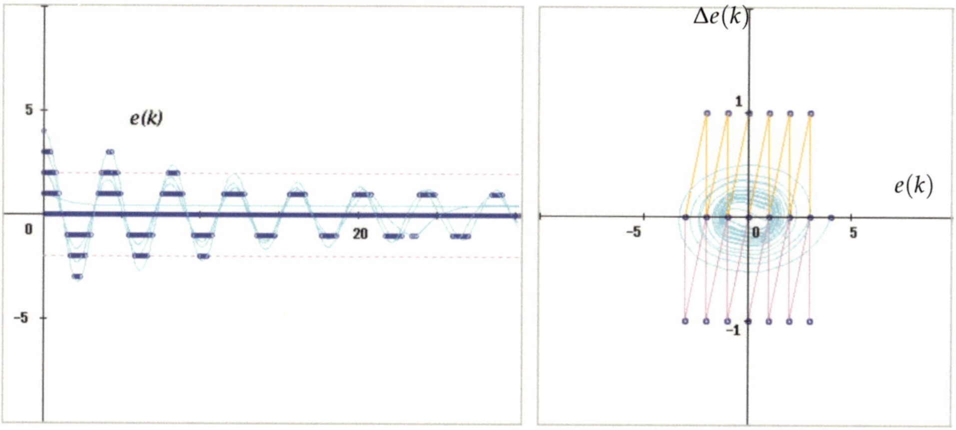

(a) (b)

Fig. 7. Time responses of $e(k)$ and $e^{\dagger}(k)$, and phase traces $(e(k), \Delta e(k))$ for Example-1 $(r = 1.0, 2.0, 3.0, 4.0, 5.0)$.

[Example-1] Consider the following controlled system:

$$G(s) = \frac{K_p(s+6)}{s(s+1)(s+2)}, \tag{45}$$

where $K_p = 1.0$. It is assumed that the discretized nonlinear characteristic (discretized sigmoid, i.e., arc tangent function (Okuyama et al., 2002b) is as shown in Fig. 6. Here, the resolution value is chosen as $\gamma = 1.0$. For C-language expression, it can be written as

$$\begin{aligned} e^{\dagger} &= \gamma * (\text{double})(\text{int})(e/\gamma), \\ v &= 0.4 * e^{\dagger} + 3.0 * \text{atan}(0.6 * e^{\dagger}), \\ v^{\dagger} &= \gamma * (\text{double})(\text{int})(v/\gamma), \end{aligned}$$

where (int) and (double) denote the conversion into an integral number (a round-down discretization) and the reconversion into a double-precision real number, respectively.

When choosing the threshold $\varepsilon = 2.0$, the sectorial area of the stepwise (point-to-point) nonlinearity for $\varepsilon \le |e| < 35.0$ can be determined as $[0.5, 1.5]$ drawn by dotted lines in the figure. In this example, the sampling period is chosen as $h = 0.1$. From (26) and (32), the max-min value can be calculated as follows:

$$\max_q \eta(q, \omega_0) = \eta(q_0, \omega_0) = 0.49,$$

when the nominal gain $K = 1.0$. Hence, $\alpha < 0.49$ and the stable area is determined as $[0.51, 1.49]$. Obviously, this sector contains the area bounded by the dotted lines. Thus, the discrete control system is stable in a global sense.

The stability condition for linear gain K can be calculated as $0 < K < 1.5$ when the sampling period is $h = 0.1$. In this example, Aizerman's conjecture for discrete-time system is satisfied. Figures 7 (a) and (b) show time responses $e(k)$, $e^{\dagger}(k)$ and phase traces $(e(k), \Delta e(k))$,

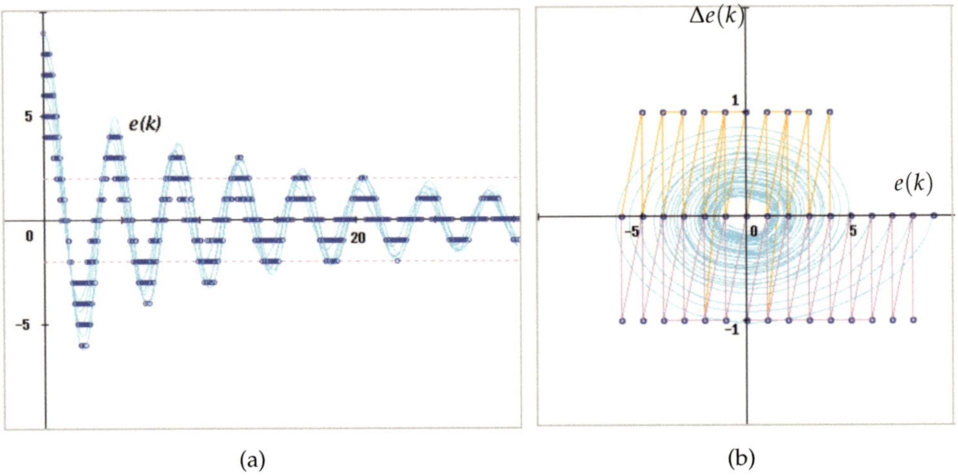

Fig. 8. Time responses of $e(k)$, $e^{\dagger}(k)$, and phase traces $(e(k), \Delta e(k))$ for Example-1 ($r = 5.0$, $6.0, 7.0, 8.0, 9.0$).

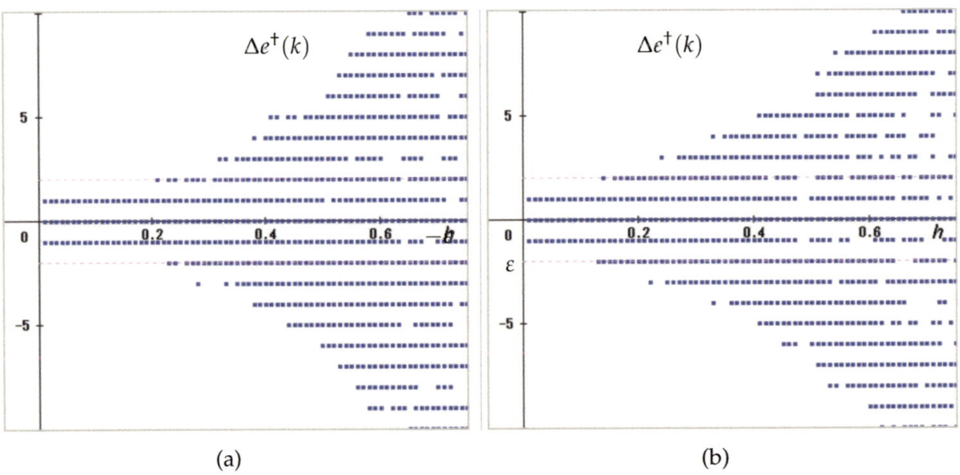

Fig. 9. Backward difference $\Delta e^{\dagger}(k)$ vs. sampling period h for Example-1 ($r = 3.0$ and $r = 9.0$).

$(e^{\dagger}(k), \Delta e^{\dagger}(k))$ of the discrete nonlinear control system when the reference inputs are $r = 1.0, 2.0, 3.0$. Figures 8 (a) and (b) show those responses when $r = 5.0, 7.0, 9.0$. Although the responses contain sustained oscillations, they do not exceed the threshold $\varepsilon = 2.0$. The input and the output of the nonlinearity lie in a parallelogram shown in Fig. 6. The robust stability in a global sense is guaranteed for all the reference inputs r. The above behavior can be estimated from the intersections of the highest gain of the sector and the stepwise nonlinear characteristic. Obviously, discrete-values $(1.0, 2.0)$ and $(-1.0, -2.0)$ lie in the outside of the stable sector.

Figures 9 (a) and (b) show the traces of backward difference $\Delta e^{\dagger}(k)$ when the sampling period h increases. As is obvious from the figure, the assumption of (10) is satisfied for $h < 0.12$ when

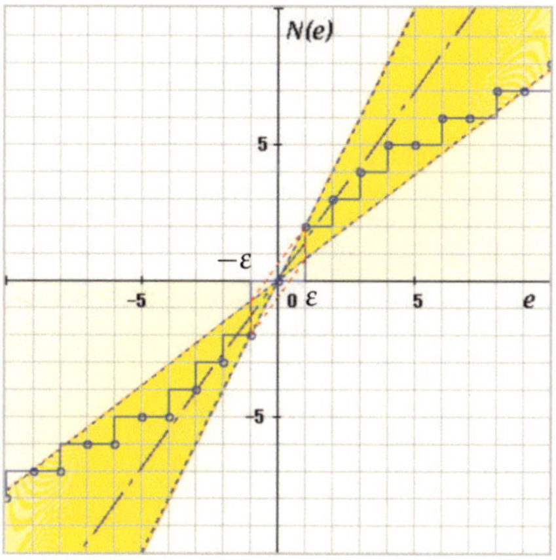

Fig. 10. Discretized nonlinear characteristic and stable sector for Example-2.

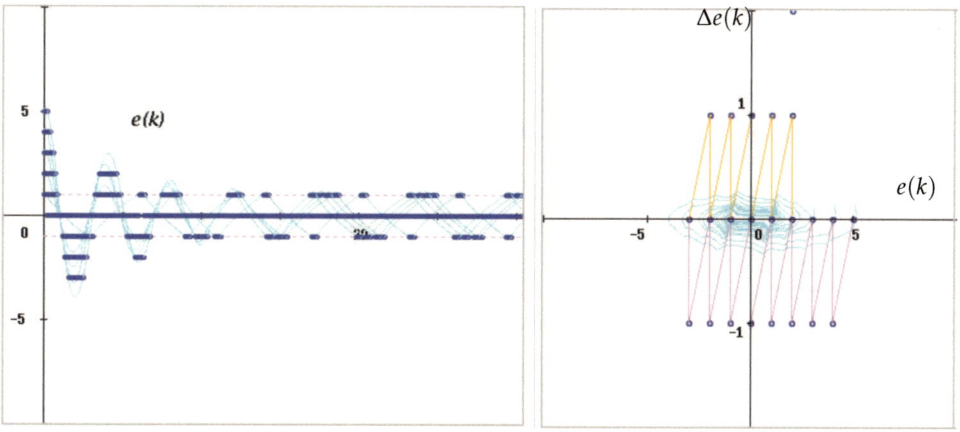

(a) (b)

Fig. 11. Time responses of $e(k)$ and $e^{\dagger}(k)$ and phase traces $(e(k), \Delta e(k))$ for Example-2 ($r = 1.0, 2.0, 3.0, 4.0, 5.0$).

$r = 9.0$, and for $h < 2.0$ when $r = 3.0$. In either case, the assumption is satisfied in regard to $h = 0.1$.

[Example-2] Consider the following controlled system:

$$G(s) = \frac{K_p(-s+8)(s+4)}{s(s+0.2)(s+16)}, \tag{46}$$

where $K_p = 1.0$. Here, the same nonlinear characteristic is chosen as shown in Example-1. When the threshold $\varepsilon = 1.0$ is specified, the sectorial area of the stepwise nonlinearity for $\varepsilon \le |e| < 10.0$ can be determined as $[0.78, 2.0]$. In this example, the sampling period is chosen as $h = 0.04$. The max-min value can be calculated as follows:

$$\max_q \eta(q, \omega_0) = \eta(q_0, \omega_0) = 0.45,$$

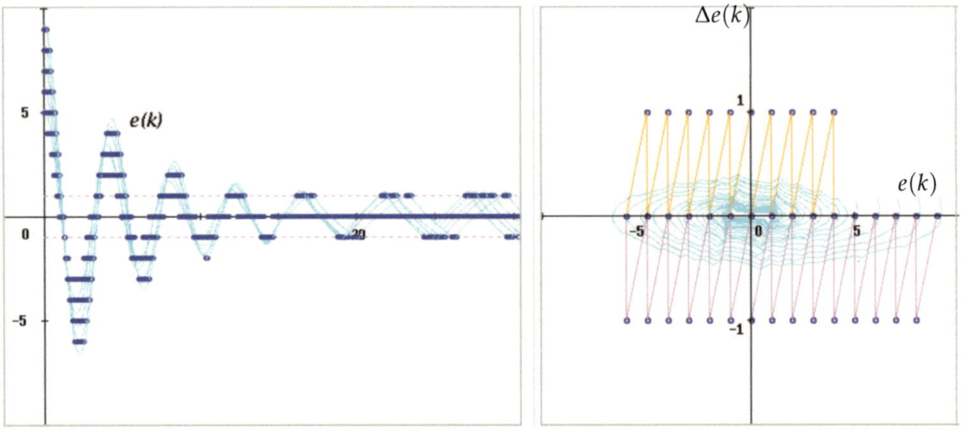

(a) (b)

Fig. 12. Time responses of $e(k)$ and $e^{\dagger}(k)$ and phase traces $(e(k), \Delta e(k))$ for Example-2 ($r = 5.0, 6.0, 7.0, 8.0, 9.0$).

when the nominal gain $K = 1.4$. Hence, $\alpha < 0.45$ and the stable area is determined as $[0.77, 2.02]$. This sector contains the above area. However, the stability region of control systems with linear gain K is given as $0 < K < 6.3$ when the sampling period is $h = 0.04$. Obviously, the discrete nonlinear control system corresponds to a counter example of the Aizerman conjecture. Figures 11 and 12 show time responses $e(k)$, $e^{\dagger}(k)$ and phase traces $(e(k), \Delta e(k))$, $(e^{\dagger}(k), \Delta e^{\dagger}(k))$ of the discrete nonlinear control system, respectively. Although the nonlinear characteristic exists in the stable area for linear systems, a sustained oscillation is generated on account of a steep build-up characteristic in the lower side of the stable sector.

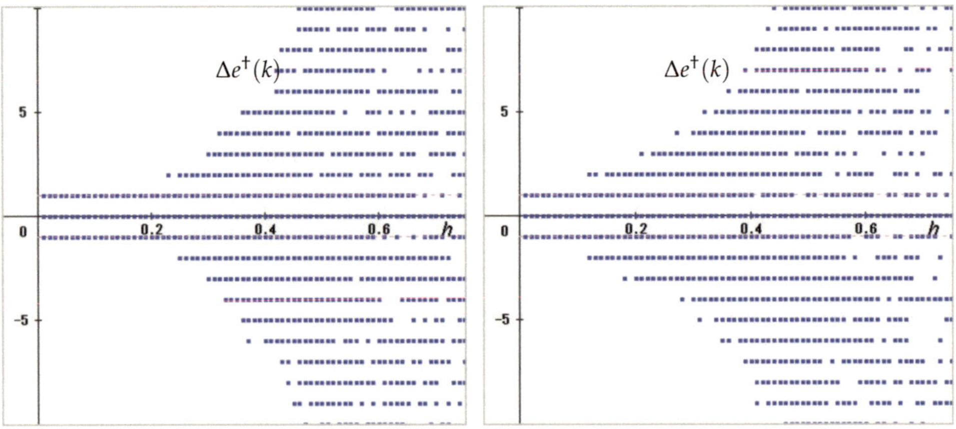

(a) (b)

Fig. 13. Backward difference $\Delta e^{\dagger}(k)$ vs. sampling period h for Example-2 ($r = 3.0$ and $r = 9.0$).

Figure 13 shows the traces of backward difference $\Delta e^{\dagger}(k)$ when the sampling period h increases. As is obvious from the figure, the assumption of (10) is satisfied for $h < 0.11$ when $r = 9.0$, and for $h < 2.2$ when $r = 3.0$. In either case, the assumption is satisfied in regard to $h = 0.04$.

10. Conclusions

This article analyzed the nonlinear phenomena and stability of discrete-time and discrete-value (discretized/quantized) control systems in a frequency domain. By partitioning the discretized nonlinear characteristic into two nonlinear sections and by defining a sectorial area over a specified threshold, the concept of the robust stability condition for nonlinear discrete-time systems was applied to the discrete nonlinear control systems. In consequence, the nonlinear phenomena of discrete control systems were clarified, and the robust stability of discrete nonlinear feedback systems was elucidated. The result described in this chapter will be useful in designing discrete (digital, event-driven, or packet transmission) control systems.

11. References

Desoer C. A. & Vidyasagar M. (1975). *Feedback System: Input-Output Properties*, Academic Press (The republication by SIAM, 2009).

Elia N. & Mitter S. K. (2001). Stabilization of Linear Systems With Limited Information, *IEEE Transaction on Automatic Control*, Vol. 46, pp. 1384-1400.

Harris C.J. & Valenca M. E. (1983). *The Stability of Input-Output Dynamical Systems*, Academic Press.

Kalman R. E. (1956). Nonlinear Aspects of Sampled-Data Control Systems, *Proc. of the Symposium on Nonlinear Circuit Analysis*, Vol. VI, pp. 273-313.

Katz, P. (1981). *Digital Control Using Microprocessor*, Prentice-Hall Internatiopnal.

Okuyama Y. et al. (1998). Robust Stability Analysis for Nonlinear Sampled-Data Control Systems and the Aizerman Conjecture, *Proceedings of the IEEE Int. Conf. on Decision and Control*, Tampa, USA, pp. 849-852.

Okuyama Y. et al. (1999). Robust Stability Evaluation for Sampled-Data Control Systems with a Sector Nonlinearity in a Gain-Phase Plane, *International Journal of Robust and Nonlinear Control*, Vol. 9, No. 1, pp. 15-32.

Okuyama, Y. (2002). Robust Stability Analysis for Sampled-Data Control Systems in a Frequency Domain, *European Journal of Control*, Vol. 8, No. 2, pp. 99-108.

Okuyama et al., Y. (2002). Amplitude Dependent Analysis and Stabilization for Nonlinear Sampled-Data Control Systems *Proceedings of the 15th IFAC World Congress*, T-Tu-M08, Barcelona, Spain.

Okuyama, Y. (2006). Robust Stability for Discretized Nonlinear Control Systems in a Global Sense, *Proceedings of the 2006 American Control Conference*, pp. 2321-2326, Minneapolis, USA.

Okuyama, Y. (2007). Nonlinear Phenomena and Stability Analysis for Discretized Control Systems, *Proceedings of International Workshop on Modern Nonlinear Theory*, pp. 350-355, Montreal, Canada.

Okuyama, Y. (2008). Robust Stabilization and PID Control for Nonlinear Discretized Systems on a Grid Pattern, *Proceedings of the 2008 American Control Conference*, pp. 4746-4751, Seattle, USA.

Okuyama, Y. (2009). Discretized PID Control and Robust Stabilization for Continuous Plants on an Integer-Grid Pattern, *Proceedings of the European Control Conference*, pp. 514-519, Budapest, Hungary.